International Conference on Maritime Innovation and Emerging Technologies 2021

Online
17 - 18 March 2021

ISBN: 978-1-7138-8301-2

Printed from e-media with permission by:

Curran Associates, Inc.
57 Morehouse Lane
Red Hook, NY 12571

Some format issues inherent in the e-media version may also appear in this print version.

Printed with permission by Curran Associates, Inc. (2024)

The Royal Institution of Naval Architects
8-9 Northumberland Street
London, WC2N 5DA
United Kingdom

Phone: 020 7235 4622
Fax: 020 7259 5912

publications@rina.org.uk

Additional copies of this publication are available from:

Curran Associates, Inc.
57 Morehouse Lane
Red Hook, NY 12571 USA
Phone: 845-758-0400
Fax: 845-758-2633
Email: curran@proceedings.com
Web: www.proceedings.com

International Conference

Maritime Innovation and Emerging Technologies

17-18 March 2021, Online Conference

PROCEEDINGS

THE ROYAL INSTITUTION OF NAVAL ARCHITECTS
8-9 Northumberland Street, London, WC2N 5DA, Telephone: +44 (0)20 7235 4622

The Royal Institution of
Naval Architects

INTERNATIONAL CONFERENCE

Maritime Innovation and Emerging Technologies

17-18 March 2021, Online Conference

Papers published in these conference Proceedings are peer reviewed.

THE ROYAL INSTITUTION OF NAVAL ARCHITECTS
8-9 Northumberland Street
London WC2N 5DA

Telephone: 020 7235 4622
Fax: 020 7259 5912

CONTENTS

THE IMPACT OF QUANTUM TECHNOLOGIES ON WARSHIP DESIGN

J Rigby, BMT, UK

SUMMARY

New advancements in quantum technologies look to provide significant improvements in capability for combat platforms in navies across the world. From communications systems to weapon targeting and navigation, there are countless possibilities for enhancements and augmentation of current systems.

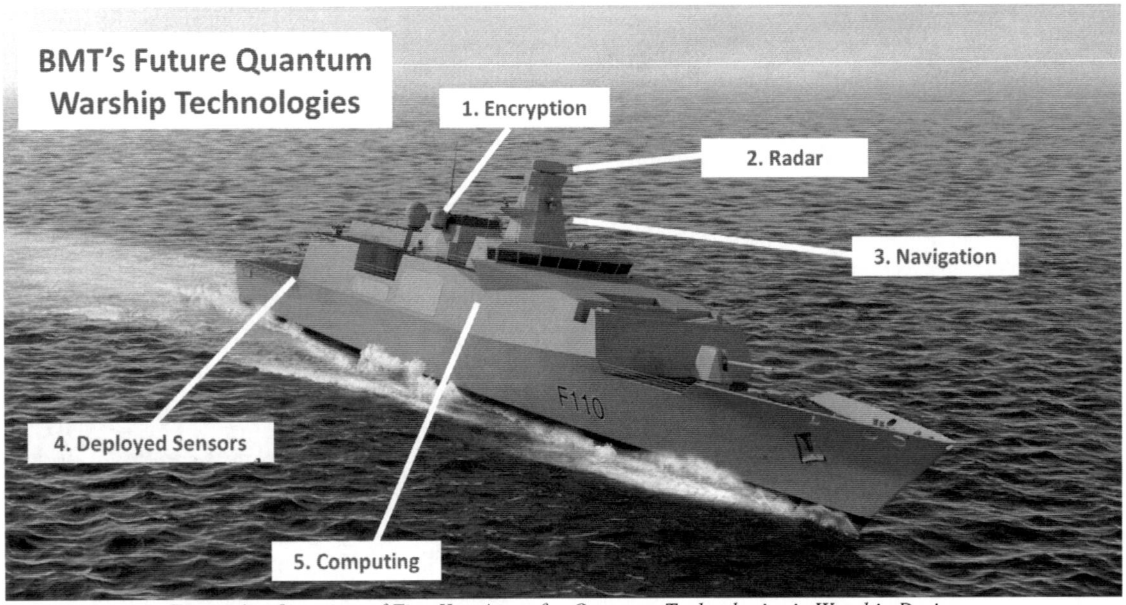

Figure 1 – Overview of Five Key Areas for Quantum Technologies in Warship Design

1. INTRODUCTION

Quantum radar and sensors offer the possibility to detect stealth platforms, above and below the water to allow more effective defence systems. Quantum encryption could provide new levels of assurance for data and voice communications providing confidence to command and control elements from strategic hubs that transmitted messages are secure. A combination of sensors and highly precise quantum clocks will enable maritime platforms to navigate with absolute confidence without the need for traditional space based global navigation satellite systems.

This paper will describe what a quantum enhanced warship of the future could look like and present the latest technological advancements from across the world. We may still be many years of the matured physical application of these technologies, but by considering the impact today we can outline the roadmap of development.

The following section outlines the key five areas for the application of quantum technologies on warships

2.1 Communications Encryption

Quantum Random Number Generation
Cryptographic algorithms often require random values exclusively articulated as numbers, known commonly as a RAND. Implementations of these algorithms need access to a Random Number Generator (RNG) which will provide random numbers when required. A Quantum Random Number Generator (QRNG) is an RNG that relies on constructions based purely on quantum mechanics for its entropy. QRNGs do not provide any new mitigation against the threat from quantum computers to traditional public key cryptography; however, they can generate random numbers at very high speed, and, the constructions produce truly unpredictable numbers.

Methods for integrating these RNGs into larger systems and assessing their behaviour are well established. In practice, the unpredictability that QRNGs can potentially offer is hard to realise. A significant reason for this is that QRNGs will necessarily sit inside classical circuitry for collection and processing, and this classical circuitry adds noise to the measurement of the quantum state.

Quantum Key Distribution (QKD)

QKD is a mechanism for agreeing encryption keys between remote parties, relying on the properties of quantum mechanics to ensure that the key has not been observed or tampered with in transit. Since traditional public key cryptography algorithms may be vulnerable to a future large-scale quantum computer, new approaches are required that do not share this vulnerability. QKD claims to offer a potential mitigation since its security properties are based on the laws of physics rather than the hardness of some underlying mathematical problems.

QKD protocols provide a mechanism for two remote parties to agree a shared secret key, where the key cannot be observed or tampered with by an adversary without alerting the original parties. However, because QKD protocols do not provide authentication, they are vulnerable to physical man-in-the-middle attacks in which an adversary can agree individual shared secret keys with two parties who believe they are communicating with each other.

2.2 Quantum RADAR

The working principles behind quantum RADAR are simple; instead of using conventional microwaves it entangles two groups of photons, which are called the 'signal' and 'idler' photons. The signal photons are sent out towards the object of interest, whilst the idler photons are measured in relative isolation, free from interference and noise. When the signal photons are reflected back, true entanglement between the signal and idler photons is lost, but a small amount of correlation survives, creating a signature or pattern that describes the existence or the absence of the target object, irrespective of the noise within the environment.

While quantum entanglement in itself is fragile in nature, quantum radar has a few advantages over conventional classical radars. For instance, at low power levels, conventional radar systems typically suffer from poor sensitivity as they have trouble distinguishing the radiation reflected by the object from naturally occurring background radiation noise. Quantum illumination offers a solution to this problem as the similarities between the signal and idler photons -- generated by quantum entanglement -- makes it more effective to distinguish the signal photons (received from the object of interest) from the noise generated within the environment. The inherent benefit of quantum radar is that it will, in theory, be able to compromise attributes that constitute stealth technologies to render them invisible to conventional radar.

2.3 Navigation

Quantum Inertial Navigation System
Inertial Navigation Systems (INS) use a range of technologies to track the true movement of a platform, but all require a positional fix both to define the initial origin of the INS and then periodically afterwards to update the filters developing the ongoing assessment of position. As errors continually build a pool of errors grows around the INS and the platforms precise position becomes less accurately known, until a new positional fix must be generated and the INS bought back to the platforms true position.

Typically these errors are of a magnitude of around 2000 yards over a 24 hour period, which means that the lay-over period between accurately establishing the platforms position is low. In open ocean a loss of Global Navigation Satellite Systems (GNSS) such as GPS for any period of time would push the conventional INS beyond the layover period and quickly impact combat effectiveness. A Quantum INS offers a revolution in accuracy and lay-over period, with the potential to practically eliminate the dependency on GNSS for own platform position and, more significantly, completely mitigate the immediate operational impact of both GNSS Spoofing and Jamming.

Figure 2 - Imperial University Quantum Compass image reproduced from Dunning (2021).

2.4 Deployed Sensors:

Navigation
While the Quantum Inertial Navigation System will provide a step change in lay-over period the performance of navigation system is enhanced when positional information can be used to supplement the performance of the INS. Using an organic Quantum Gravimeter the platform will be able to conduct Simultaneous Location And Mapping (SLAM) as it moves through its operating areas and maintain a constantly updating assessment of its true position. The use of both Quantum capabilities in tandem will make the navigation system independent of GNSS and enable it to operate indefinitely without access to external positioning signals.

Anti-Submarine Warfare
Quantum Gravity Sensors are capable of detecting even very small changes in gravity, and therefore the presence of large submerged objects, with the potential to revolutionise the detection and localisation of submarines. With evolving TRL and stability Quantum Gravimeters will become viable in Maritime Patrol Aircraft, followed by the development of lighter sensors

which could give platforms the opportunity to deploy both autonomous and conventional gravity sensing drones to sanitise the water ahead of a moving Task Group or conduct Anti-Submarine operations over wide areas.

2.5 Quantum Computing:

These devices exploit the complex nature of quantum information. While a conventional computer uses binary (bits), a quantum computer uses qubits. Each qubit can be zero, one, or both meaning it can be put into a quantum superposition. Because qubits can be put into many states at once, a quantum computer can process many inputs simultaneously instead of having to go through them one by one like a conventional machine. For some types of problem, this can mean a much faster solution.

For this reason quantum machines are much faster at searching large databases and factoring large numbers which is of critical importance because it is behind the most common form of cryptography, which is used to protect financial and other sensitive data. There are other applications in design and modelling that would vastly improve and expedite current processes, including aircraft, ship, boat and weapon system design.

3. CONCLUSIONS

In conclusion, there are huge opportunities for the application of quantum technologies for use in commercial shipping and warships. However, on the whole the technologies are still quite immature, currently confined to lab based demonstrations and a long way off full implementation.

One quote that sticks with me is from Richard Feynman, an American Quantum Physicist from the 20th century, talking about the paradox of quantum physics.

"The 'paradox' is only a conflict between reality and your feeling of what reality 'ought to be'. If you are not completely confused by quantum mechanics, you do not understand it"- Richard Feynman

The above quote explains how this is not a simple topic and there is still a lot that needs to be done to understand Quantum Mechanics and the tools required to capture its sensing power.

Some of the overarching key challenges include:

- **Economic quantum particle capture and cryogenics** – Many of the current lab based systems require cryogenic temperatures close to absolute zero to enable the effective capture and tracking of a particle. This needs to become economical or find a way to increase overall temperatures.
- **Sensitivity and Focus** – Some sensor systems that have got working lab based protoypes can also be too sensitive and overwhelmed with information. We need to find a way to channel and focus the detection power to make the outputs useful.
- **Reducing Qubit Error** - Manipulating individual qubits can introduce errors in calculations, and unless that error rate falls below a certain allowable level, then entangling more and more qubits will only add more noise to the system. We need to find a way to reliably upscale from singular qubit systems.

Although these challenges are large, the rewards are equally huge. If these and other challenges can be solved then it will signal the biggest technological leap in sensing and computational technology the world has ever seen.

4. ACKNOWLEDGEMENTS

The authors would like to acknowledge all those who supported the quantum technologies research conducted at BMT and the production of this paper.

5. REFERENCES

1. Dunning, H., 2021. Quantum 'compass' could allow navigation without relying on satellites | Imperial News | Imperial College London. [online] Imperial News. Available at: <https://www.imperial.ac.uk/news/188973/quantum-compass-could-allow-navigation-without/> [Accessed 14 February 2021].

2. DWave, 2021. Quantum Computing Primer. [online] Dwavesys.com. Available at: <https://www.dwavesys.com/tutorials/background-reading-series/quantum-computing-primer> [Accessed 14 February 2021].

3. Feynman, R., 2021. Richard P. Feynman Quotes (Author of Surely You're Joking, Mr. Feynman!). [online] Goodreads.com. Available at: <https://www.goodreads.com/author/quotes/1429989.Richard_P_Feynman> [Accessed 14 February 2021].

AUGMENTED REALITY INTEGRATION PLATFORM FOR SHIP NAVIGATION

D Bandara, M Woodward, C Chin, D Freeman Australian Maritime College, University of Tasmania, Australia

SUMMARY

Holograms are the subject of sci-fi but, like many ideas once deemed impossible, fiction and reality are converging. As new technology reduces in cost, opportunities arise in the maritime sector. This research strives to make objects that may be hazardous for navigation more easily seen in poor visibility. The wider study aims to allow the crew to see holograms of the navigation-lights of other vessels and chart-objects when they would be otherwise obscured by fog. This needs the bringing together of a variety of real-time data from bridge equipment. Each device has a different protocol/dated format, that in some cases may present conflicting information. The research presented here addresses the problem of reviewing data sources, categorising data types/strings, and converting them into a common language; to define/locate the holograms. The collected information is processed using bespoke algorithms. The outcomes present the selected methodologies and derived solutions for the holographic navigational enhancements.

NOMENCLATURE

AIS	Automatic Identification System
AR	Augmented Reality
ECDIS	Electronic Chart Display and Information Systems
ENC	Electronic Navigation Charts (ENC)
GPS	Global Positioning System
HMD	Head Mounted Device
HUD	Head Up Device
IMO	International Maritime Organization
NMEA	The National Marine Electronics Association
VR	Virtual Reality

1. INTRODUCTION

Imagine if your ship is in the process of manoeuvring on a very foggy day; so that you will not be able to see the surrounding environment through ship bridge windows. The bridge is a hive of activity, with the helmsperson controlling the ship, the Master providing verbal commands and various other crew undertaking their supporting roles, every person standing in the bridge is very busy looking at the various displays of navigation aids, aiming to understand and sort out the information available to them. They are doing all the necessary calculations to path planning and to make sure the safe navigation. Now imagine what it would be like if the helmsperson, the Master and the supporting staff were all wearing Augmented Reality (AR) headsets. All have a normal view of their immediate local surroundings, so they should not bump into each other or trip over the bridge equipment. Externally to the bridge however they each share the same enhanced holographic view. Thus, they are able to carry out their duties unimpeded while at the same time having an enhanced perspective. From their perspective, the fog has become, for all intent, transparent. The helmsperson, Master and supporting persons unanimously share an enhanced experience of reality; persevering the location of nearby collision hazards via the AR headsets.

It goes without saying that this would be quite astonishing in ship operation in the described compromised visibility conditions. The question remains however, how can we bring all the navigation information into the head-mounted AR device? To start with, what information is readily available, how is it transmitted or extracted for use and in what format is it delivered. Then, how do we organise this data in a practical way and in real time?

The Latin word "Augere" has the meaning of "to increase" or "to add", which is used to create the word "Augmented". Virtual reality (VR) visually immerses the user totally in a digital view. As illustrated in Figure 1, the user can still fully see the real world around them as well as the holographic object; in this case a tug boat. New technologies that can be utilized to present information in a holographic form are reaching a state of technology readiness for possible application in supporting maritime operations and training. Therefore, it is proposed as important that investigations be made to determine how AR may be employed to provide a complementary technology for enhanced safety in maritime operations in compromised visibility. It is important to consider the information provided to the bridge crew and the way it is provided to ensure it is making best use of available and emerging technologies.

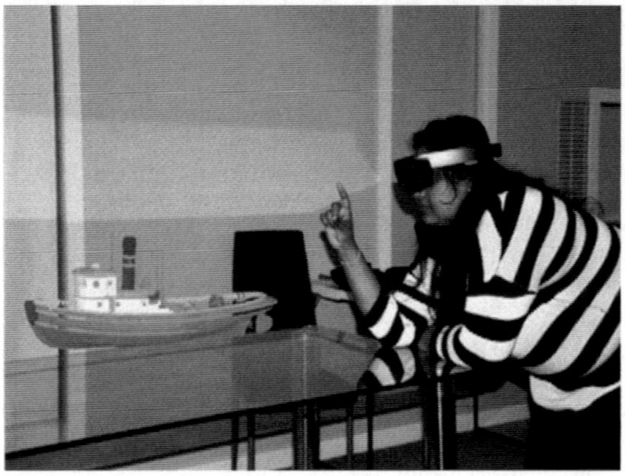

Figure 1. The user sees the real world as well as virtual world

54% of maritime accidents in 2018 were due to collisions, contacts and groundings; making it extremely important to understand the factors that cause these accidents [1]. According to International Maritime Organization (IMO) 2010, 90% of maritime accidents due to collisions at sea are caused in some way by human error [2-4]. It is noted that the quality of information provided to the bridge crew plays an important role. Bridge crew are often exposed to large amount of information and that this information is not well organized for them to use. This results in information overload and produces a 'big data' problem. To address this problem, this research study uses AR as a platform to develop the integrated and interactive user interface to enhance visualization for ship operations in compromised visibility conditions such as snow, fog, rain, and rough seas. This study investigates how to import data such as Global Positioning System (GPS), Automatic Identification System (AIS) and Electronic Navigation Charts (ENC) coming from a range of sources and delivered in a range of varying protocols. The main contribution of this part of the research study is to focus on harmonising the identified data into a common platform. Using this platform, a holographic interface will later be developed, with the intention of presenting the data in an integrated, clear and configurable form. Verification of the functionality of the proposed integration of data is presented through case studies for potential application.

2. RESEARCH BACKGROUND

The history of AR prototypes begins in the 1960s as a show of 3D graphics [5]. Since then, use of AR technology has increased rapidly in various industries such as training and education, medical, engineering, architecture, defence operations [6-10]. The commonly used Head-Up-Display (HUD) AR device is employed in fighter jets and commercial aircrafts to reduce their training time and enhanced the ability of pilots [11]. With recent advances in computing power and display technologies, new Head Mounted Devices (HMD) are available in the market. There are companies such as Google, Samsung, Epson, Sony and Apple who introduced their own immersive wearable devices. The descriptive literature review of the AR technology can be found in the authors previous paper [12]. This paper pointed out that, the Microsoft HoloLens was ultimately selected as the AR device based on both its functionality and on account of being readily available. The first-generation Microsoft HoloLens was used for this study and full specification of the device is available in [13]. As per the reported previous studies, there are few studies that have conducted research into potential use of AR in the maritime industry.

Holder and Pecota [14] claimed that the 'head-down' position time is not enough when using electronic navigation aids and that caused seafarers to be reluctant to use navigation aids. These results showed that it is important to display standard information in a more user friendly way across situations that could include other tasks and vessel specific information. Another research team integrated a visual system with a thermal camera as well as augmentation and found that the functionalities provided by AR must differ according to people and weather conditions, requiring the provision of contextual information [15]. Oh, Park [16] investigated AR based advanced navigation aids systems, implementing the representation of information using AR. They found that while various types of high-tech navigation equipment are available to improve the operational safety of vessels, further research is needed on the method of displaying and servicing information efficiently rather than providing a non-discriminatory information grant.

AR should already be able to project information from display devices in bridge equipment to appear as a head-up display. Using a more advanced and contemporary application, it is proposed that AR can also be used to create holographic representations of real or digital (Holographic) objects to enhance situational awareness. Complex information from multiple sources coming from navigation aids in ship bridge may be presented to the bridge crew in a more intuitive way that may reasonably be suggested to lead to more safety and efficient operations at sea. This research paper focused on how to integrate all the information available at ship bridge into a common digital platform; that will later inform the AR interface.

2.1 DECONSTRCTION OF THE TECHNICAL PROBLEM

New technologies that can be applied for safer ship navigation are appearing on the market all the time. Legislative bodies, such as the IMO and national Maritime Safety Administrations (MSAs), publish and maintained extensive rules and regulations on safety standards on the safe operations at sea. Both aspects strive to address the safety of passengers and vessel, which is one of the major concerns in maritime industry. Nevertheless, the number of maritime accidents has not reduced as expected [17-

22]; with most maritime accidents reportedly due to the collisions, contacts, and grounding. Many studies focus on how humans interact with technology and interpret and interact with information provided. This is not within the scope of this study, but is understood to be a driving factor. As a practical engineering research study, this work focuses on the application of technology. The surrounding study had identified that navigation lights are a practical way of presenting hazards to the ship's crew. This conclusion is firstly found from a practical point of view for minimising data processing times; lights being vastly simpler to project than whole objects such as ships. Secondly, this conclusion is drawn from recognising that the ship's crew are perfectly familiar with navigation lights and the information that they present.
Based on the above, the main problems at hand are:

- What digital information is typically available on the ships bridge?
- What format or protocol is that data held in?
- How is the data accessed/transmitted?
- How can the necessary aspects of any given data source be isolated and extracted for further processing?
- What mathematical manipulations are needed to convert the data into a usable/compatible form?
- What are the sensitivities of various conversion methods on the estimated location of target objects?

Maritime personnel working on the bridge of a ship are presented with a multitude of information, from a range of sources; that may include GPS data, ENC data, and AIS. Each of these information sources are presented in a different way or form, placing a certain cognitive demand on those using it. This study proposes to harmonise into a common platform and present the information in a more uniformly organised, and contemporary way. For example, it is proposed that, instead of having to interpret the multitude of information, to ascertain the location of another nearby vessel (in poor visibility), that a hologram of that ship's navigation light be visible to the crew at exactly the same location as the real ship in real-time. To achieve this, the above mentioned problems need addressing to extract the relevant and necessary information from existing systems and process for onward application.

3. INTEGRATION PLATFORM

To derive the data integration platform, a series of steps need to be addressed. The first task is to identify what data is available on the ship bridge. This may include understanding what equipment is available, how it works and/or what it strives to offer the seafarer, specifically in terms of collision avoidance. To achieve this an audit of typical bridge equipment is made and the information categorised. At this stage it is also necessary to identify exactly how the data is obtained. For example, Wi-Fi or verbal communication or even just looking out of the window. For many applications, digital data is collected and transmitted in a standardised and agreed format through the Pilot Port. This plug-in application is available on many ships; details given in [17].

It is important to investigate ways of projecting holographic objects such as boat/ship, light house, fixed markers, buoys so that they can appear when viewed from the bridge of a real ship. Figure 2 shows the information flow process that will be used. Starting top left, a database of common objects is defined, giving normalized geometric light configurations for various ships types and chart objects. This is further described in Section 3.3.

Next the GPS is taken to provide data about the position of our ship and/or the position of other vessels via AIS. This data comes through the Pilot Port as an NMEA string (see Section 3.2). This string contains the vessel position in terms of latitude and longitude in the Geographical Coordinate Systems (GCS). Similarly, AIS, which is also a NMEA string is used to extract information about the position, heading/orientation, ship length, and ship type. This is expanded upon in Section 3.1. The Electronic Chart Display and Information System (ECDIS) (or ENC) is used to obtain information about lit chart objects. This includes details on the position and height of any light, the light colours of any flashing pattern.

Each of the above outputs data in its own unique format and protocol, that must be interpreted and converted to a common form. The format and processes used are explained in Section 3.2.

Using the processed information collected from the GPS/AIS and ECDIS, appropriate matching objects can be selected from the derived database. Each object is then scaled to size/orientation according to the desired values. In parallel, the position of objects with respect to 'our own ship' frame-of-reference is calculated; with details given in Section 3.4.

Combining all of the above, the integration platform is then ready to output information to the Augmented Reality device describing what to display and where and how to display it.

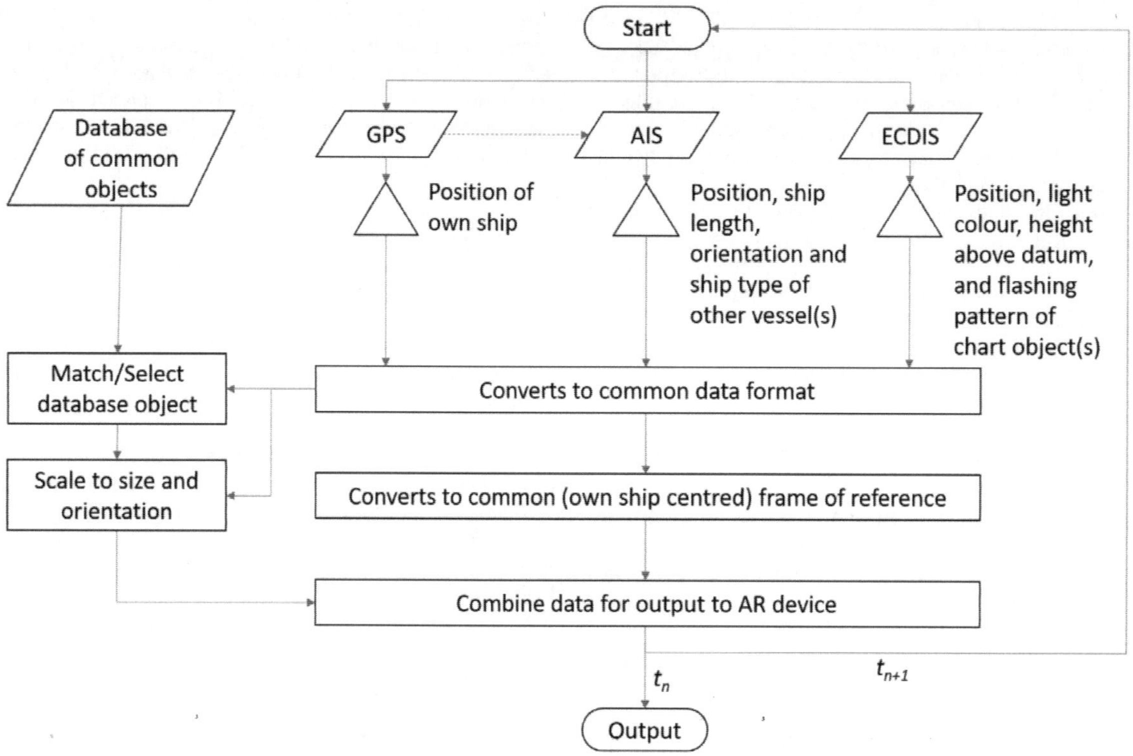

Figure 2. Conceptual Design of AR Integration Platform

3.1 REVIEW OF DATA SOURCES

There are multiple sources of data available in the ship bridge in the form of equipment, logbooks and documentation (procedures and policies). It is therefore important to identify the usage of these information resources in different maritime operational modes. We first identify the primary information needed for the ship navigation such as vessel's own speed, buoys and shipping lanes and international COLREGs rules and other critical information when it comes to maritime navigational safety. Next, we identify the secondary information such as information on shallow waters, underwater obstacles, wrecks (this information can be obtained from ECDIS), information on other vessels, GPS information and radar information. In this study, we will focus on GPS, AIS and ENC data types.

Global Positioning System (GPS) provides digital information on position, in the form of longitude and latitude that is output in ASCII form as a standardised data string. This can be used for both our own location and the location of target object via an AIS signal.

Automatic Identification System (AIS) is used to obtain data about other vessels via a data link delivered in an ASCII code standard string. This string contains information about position (from GPS). In addition, other data utilised in this study includes: heading, ship-length, and light configuration (selected from the database) based on ship-type.

Electronic Navigation Chart (ENC) is a software program that incorporates GPS signal and other sensor data to display the vessel's moving position on a digital chart, along with parameters such as course over ground (COG) and speed over ground (SOG). There are two broad categories of digital charts called raster charts and vector charts used in ENC.

The study first reviews and critiques all sources of information provided to the bridge crew for navigational operations. The data is deconstructed with the specific intention of isolating and extracting only the information relevant for mitigating collision risk in compromised visibility. As this data will necessarily come from a range of sources, delivered in a range of varying protocol, the research focuses on harmonising the identified data into a common platform as illustrated in Table 1. Using this data/platform, a holographic interface will be developed, with the intention of presenting the data in an integrated, clear and configurable form. Finally, verification of the functionality of the developed interface will be undertaken.

3.2 PROTOCOL OF DATA TYPES

It is important to understand the raw data format of primary and secondary information needed for ship navigation. Moreover, there is a need to know how to communicate these data in raw data or string format and the way this information re-presents in current navigation aids. The NMEA (The National Marine Electronics Association) defines an electrical interface and data protocol called 'NMEA 0183 or NMEA 2000' for communication between marine instruments [18, 19].

When it comes to GPS, the NMEA defined a RS-232 communication standard for devices that include GPS receivers. The GPS receivers can output geospatial location, time, headings and navigation-relevant information in the form of ASCII comma-delimited message strings. ASCII stands for American Standard Code for Information Interchange. ASCII code is the numerical representation of a character such as 'a' or '@' or an action of some sort. The NMEA standard dictates how each string is formed with a dollar sign ($) leading each new GPS message [20].

An example of the string output from a GPS receiver is:

$GPGLL,3723.2475,N,12158.3416,W,161229.487,A,A*41

AIS is another information source of information available for navigation, in which transmitters are fitted to vessels, navigation markers, and certain types of shore station. They periodically squawk their position (and course, when applicable) using TDMA (Time Division Multiple Access) technology similar to the way cell phones do to avoid mutual interference. AIS receivers make this data available for navigation, anti-collision systems, and other uses. AIS receivers report ASCII data packets as a byte stream over serial or USB lines, using the NMEA 0183 or NMEA 2000 data formats. The RS422 variant of serial specified as a physical layer by NMEA 0183 is common in marine navigation systems; there may be a "pilot plug" which converts to USB. Alternatively, newer AIS receivers may report directly over RS232 or USB. AIS packets have the introducer "!AIVDM" or "!AIVDO"; AIVDM packets are reports from other ships and AIVDO packets are reports from your own ship [21].

An example of the string output from a AIS is:

!AIVDM,1,1,,B,177KQJ5000G?tO`K>RA1wUbN0TKH,0*5C

Table 1. Integration platform for Bridge Data Process

Navigation aid type:	**GPS**
Information:	**Data Process:**
Latitude and Longitude	1. Receive satellite signals to GPS antenna 2. Software Process these data into a digital format 3. Display in digital format (ASCII) in monitors 4. User will examine these positions and make interpretation based on training and experience 5. System software will process these positions 6. System will display the latitude and longitude in pre-defined format
Transmutation:	
NMEA data (Pilot Port)	
Leads to know:	
Own vessel Position relative to world coordinates	
Navigation aid type:	**ENC (Defined object – Lighthouse)**
Information:	**Data Process:**
Position of target	1. ECS process the relevant charts from the data base. 2. ECS display the information 3. User will examine visual output and make interpretation based on training and experience 4. User will identify known and unknown area information such as land area, coastal lines, buoys, trees, hazards marks, etc 5. User will select target land area (lighthouse) from defined symbols on ENC using cursor (and click it) 6. User will select functions such as zooming, highlighting and etc (and click on it) 7. System software will calculate the position of lighthouse from the selected cursor location 8. System will display calculated position of lighthouse in pre-defined format 9. User will read the position of lighthouse
Transmutation:	
NMEA data (Pilot Port)	
Leads to know:	
Position of other vessels relative to fixed or moving objects	

Navigation aid type:	**ENC (un-defined object – Big Tree)**
Information:	**Data Process:**
Position of target **Transmutation:** NMEA data (Pilot Port) **Leads to know:** Position of other vessels relative to fixed or moving objects	1. ECS process the relevant charts from the data base. 2. ECS display the information 3. User will examine visual output and make interpretation based on training and experience 4. User will identify known and unknown area information such as land area, coastal lines, buoys, trees, hazards marks, etc 5. User will select other objects near the target area and highlight that area 6. User can look into RADAR data and get decision on selected area 7. User will select target land area (Big Tree) from defined symbols or without symbols on ENC using cursor (and click it) 8. User will select functions such as zooming, highlighting and etc (and click on it) 9. System software will calculate the position of the tree from selected cursor location 10. System will display calculated position of the tree in pre-defined format 11. User will read the position of the big tree
Navigation aid type:	**AIS (Other vessel)**
Information:	**Data Process:**
Dynamic data (latitude, longitude, position accuracy time, course, speed, navigation status) Static data (name, dimensions, type, draft, destination and estimated time of arrival) **Transmutation:** NMEA data (Pilot Port) **Leads to know:** Position of other vessels relative to fixed or moving objects **Transmutation:** **Leads to know:** Heading with respect to the true north	1. Transmitter and receivers must be on status for both vessels 2. Data have to process using relevant software 3. These data display on monitors 4. User will examine these data output and make interpretation based on training and experience 5. User will identify known target information (Dynamic and Static data) display on the screen 6. User will select target using cursor (and click it) 7. System will display pre-defined format of the information of selected vessels 8. User will understand the relevant information and get opportunity to contact the relevant vessel using VHF communication

3.3 CONCEPTUAL DESIGN OF THE INTEGRATED PLATFORM

A previous study by the authors has determined that holographic navigation lights can visually replicate existing navigation lights; when taking into account distances over water comparable to those in real navigation cases [22]. The paper demonstrated the feasibility of projecting holographic replications of real-world objects in the same geo-location. The study derived and calibrated a scaling algorithm needed to deliver realistic holographic navigation lights when projected over water and over suitably long distances. This led to, as outlined in Figure 3, the derivation of a database of standard objects that, when scaled correctly, present a realistic representations of navigation lights. The database is formulated on tabulated data and gives normalised dimensional light configurations. Applicable to both ships and chart objects, each table contains information about six possible lights: port; starboard; stern and three mast head lights.

Table 2 gives by example, four cases: A ship restricted in its ability to manoeuvre; a large vessel underway; a lighthouse; a channel marker. The database is derived with a selection of representative light configurations based on COLREG's. The configuration is based on a box of unit dimensions with the centre of the base assumed to be centred on amidships or the mid-section of the object such as a lighthouse.

Longitudinal, transverse and vertical dimensions of the object lights are then given as unit ratios, with a left-handed Cartesian coordinate-system, such that they will scale appropriately when multiplied by the representative length of the target object/ship. In column five, a number is defined to give the expected colour of the light. That is, 1 for red, 2 for green and 0 for white. In column six a data string is derived describing the angle through which each light is visible; the first number giving the starting angle and the second the closing angle, both in degrees with zero being directly ahead. The last column provides a data string as a control for flashing lights such as with a lighthouse. The first number giving the number of flashes and the second giving the length of time between groups of flashes.

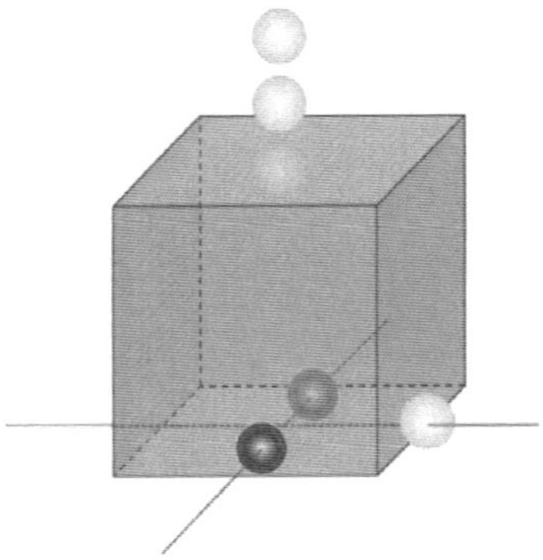

Figure 3. Representation of vessel in terms of lights

Table 2. Database for standard dimension

Vessel Restricted in ability manoeuvre						
Item	**x - offset (-)**	**y - offset**	**z - offset**	**Colour number**	**Angle (degrees)**	**Pattern**
Port	0	-0.2	0	1	247.5;360	0
Stbd	0	0.2	0	2	0;112.5	0
Stern	-0.5	0	0	0	112.5;247.5	0
MHLI	0	0	-0.2	2	247.5; 112.5	0
MHL2	0	0	-0.25	0	247.5; 112.5	0
MHL3	0	0	-0.3	2	247.5; 112.5	0
Large Vessel (Underway)						
Item	**x - offset (-)**	**y - offset**	**z - offset**	**Colour number**	**Angle (degrees)**	**Pattern**
Port	0	-0.2	0	1	0	0
Stbd	0	0.2	0	2	0	0
Stern	-0.5	0	0	0	0	0
MHLI	0	0	-0.2	0	0	0
MHL2	0	0	-0.25	0	0	0
MHL3	-	-	-	-	-	-
Lighthouse						
Item	**x - offset (-)**	**y - offset**	**z - offset**	**Colour number**	**Angle (degrees)**	**Pattern**
Port	-	-	-	-	-	-
Stbd	-	-	-	-	-	-
Stern	-	-	-	-	-	-
MHLI	0	0	1	0	0;360	2;3
MHL2	-	-	-	-	-	-
MHL3	-	-	-	-	-	-
Port Channel Marker						
Item	**x - offset (-)**	**y - offset**	**z - offset**	**Colour number**	**Angle (degrees)**	**Pattern**
Port	-	-	-	-	-	-
Stbd	-	-	-	-	-	-
Stern	-	-	-	-	-	-
MHLI	0	0	0	0	0;360	0
MHL2	-	-	-	-	-	-
MHL3	-	-	-	-	-	-

3.4 FRAME OF REFERENCES

It is important to understand how to define an object of interest in its direction and distance in relation to a 'local' coordinate system. The problem to be addressed is in knowing the best way to do that when moving between a 2D and 3D reference frame. That is, when acknowledging that the real world is on a curved surface, how to represent things in a flat surface frame-of-reference.

To do the necessary calculations, the physical horizon at sea-level is assumed to be 18 miles (29km). There are of course existing methods that can be used in this simple trigonometric calculation. It is however necessary to demonstrate and understand the problem and assure that the chosen method will deliver credible results within the magnitude of values that are of interest in this problem. Therefore, considering two angles related to longitude and latitude will uniquely define the position on the surface of any arbitrary shape. The distance from the axis reference to the intersection with the surface is the only measurement that will be affected by the difference in shape. The effect of elevation estimate will be on the precision of quoted distance of an object from the local location as 'A' and target vessel position as 'B'; where the distance from A to B is assumed to be 29 km; which is denoted by Z_3 as shown in Figure 4.

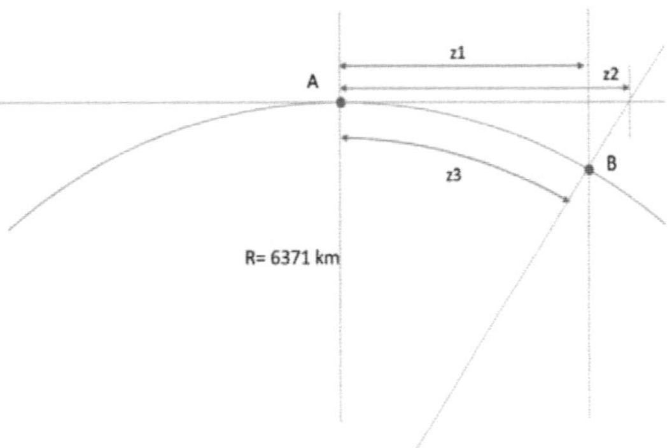

Figure 4. Vessel position on the earth

The earth is considered as a sphere with a mean radius of 6,371 km. From basic trigonometry Z_1 and Z_2 can be calculated. The length of the arc depends on the radius of a circle and the central angle θ. We know that for an angle equal to 360 degrees (2π), the arc length is equal to the circumference. From this explanation arc length Z_3 is:

$$Z_3 = \frac{\theta}{180} \pi R$$

Substituting Z_3 and R values,

$$\theta = 0.261^0$$

Therefore, $Z_1 = R \, Sin \, \theta$ and $Z_2 = R \, tan \, \theta$

$Z_1 = 28.9999$ km and $Z_2 = 29.022$ km

Considering the other factors affecting the elevation position (distance from the centre of the earth) of the vessel values are assumed: +/- 6.1 m changes in tide, +/- 0.63 m air pressure changes to water surface and +/- 1.5 m due to sea-state. However, taking the 99% confidence interval for the root-sum-square of the sensitivity and uncertainty of the radius the difference in Z_1 is found to be 2.75×10^{-6}. In terms of % change this is in the 11th decimal place; and so is negligible. Repeating the same calculation with the target ship position 50m away, the difference in 'Z' values is not significant and the uncertainty drops to $\times 10^{-14}$. Thus, for this project, the projection error is reasonably neglected as it would not be discernible with the naked eye.

4. RESULTS AND DISCUSSION

The overall aim of the study is to explore the viability of projecting augmented reality navigation of collision hazards; such as other ships and lit chart objects. The aim is to make these objects clearly visible when they would not otherwise be so; such as in fog. The objective is to provide information to the bridge crew in an intuitive way and in a form that they are familiar with. In addressing the last point, navigation lights are selected as a 'language' that seafarers are assumed to be fully aware of. Then, the aim of this paper is to present the derivation of an integration platform; that is a necessary part of the wider study. The integration platform must extract data from a number of sources and in a number of different formats and convert them to a useable form. To this end, this paper has presented the detailed deconstruction of the technical problem, and the process flow used to address the needs.

Based on previously published work, navigation lights are selected as the common language, being both commonly understood (familiar to seafarers) and computationally efficient (less demanding to display than an image of the whole ship).

A database methodology is derived and presented that stores a selection of standard objects. This includes various ship light configurations and lit chart objects such as lighthouses and channel markers.

Data sources are identified and categorised. In each case, the data protocol and format are identified; such that practical extraction is realisable. In particular, the examples of GPS, AIS and ECDIS delineated within the paper.

Next, a data processing algorithm is derived and presented that takes the raw data and, combining it with the database objects, outputs the necessary information for enacting the Augmented Reality projections.

Overall, as a part of a wider study, the paper addresses a 'big data' problem; related to the provision of hazard information aboard ships. The wider aims of the global study strive to address issues of collision avoidance by providing information in a more intuitive way. This is nevertheless assumed implicit, driven by data/information reduction rather than addition. That is, the scope of the study does not explicitly investigate the performance of seafarers. Rather, the study makes the assumption that, by simplifying information and presenting it in an already familiar form, that it will result in reduced cognitive load. This by no means infers that this is the case and, as the technology develops, it is hoped that specialist in this area will investigate such assumptions.

5. CONCLUSIONS

The specific aim of this study is to deconstruct the technical problem of a uniquely defined technical issue. Namely, to identify, what information is available, in what format, and how to extract the data, necessary to make the desired holographic projections of the navigation lights of other vessels and chart objects. The objective is to derive an Integration Platform algorithm that address the need of the task. As such the paper presents a deconstruction of the technical problem leading to the derivation of an object database and a methodology for extracting and processing the necessary information for onward use. The derived functionality identifies that bridge equipment including GPS, AIS and ECDIS, when combined in the prescribed way with the derived database objects, provide the complete and necessary information. The proposed algorithm/methodology meets the objectives in that, given the required inputs, it provides the required output needed.

6. FUTURE WORKS

The next stage of the study is to actualize the Augmented Reality projections. Future work will conduct real-time case-study experiments using AMC surface vehicle or Full-Mission Ship Bridge Simulator (FMSBS) to verify the functionality of the proposed AR platform. Verification and validation of these functionality will be carried out at a technical level. Furthermore, technical, audio-visual outputs and processor limitations of HoloLens when running the developed applications will be explored.

7. ACKNOWLEDGEMENTS

We would like to thank all the staff members in the bridge simulator, AMC and University of Tasmania.

8. REFERENCES

1. MAIB, *Annual Overview of Marine Casualties and Incidents 2018*. 2018: Southamton, UK.
2. Acar, U., R. Ziarati, and M. Ziarati, *Collisions and groundings–major causes of accidents at sea*. An investigation into COLREGs and their application at sea. 2012, Pori, Finland: Satakunta University of Applied Sciences. 40-47.
3. Gauss, B., M. Rötting, and D. Kersandt. *Naridas–evaluation of a risk assessment system for the ship's bridge*. in *Human Factors in Shop Design, Safety and Operation. RINA-The Royal Institution of Naval Architects. International Conference*. 2007.
4. Squire, D., *The human element in shipping*. Retrieved July, 2005. **1**: p. 2008.
5. Van Krevelen, D. and R. Poelman, *Augmented reality: Technologies, applications, and limitations Vrije Univ.* Amsterdam, Dep. Comput. Sci, 2007.
6. Akçayır, M. and G. Akçayır, *Advantages and challenges associated with augmented reality for education: A systematic review of the literature.* Educational Research Review, 2017. **20**: p. 1-11.
7. Guo, P. and D. Kaplan, *Augmented reality based component replacement and maintenance*. 2019, Google Patents US 10 , 198 , 620 B2. p. 18.
8. Pratt, P., et al., *Through the HoloLens™ looking glass: augmented reality for extremity reconstruction surgery using 3D vascular models with perforating vessels.* European radiology experimental, 2018. **2**(1): p. 2.
9. Slocombe, G., *Augmented reality is a useful technique for defence.* Asia-Pacific Defence Reporter (2002), 2017. **43**(8): p. 94.
10. Wang, W., et al., *Holo3DGIS: Leveraging Microsoft HoloLens in 3D Geographic Information.* ISPRS International Journal of Geo-Information, 2018. **7**(2): p. 60.
11. Kim, G.-H., S.-I. Jo, and J.-H. Park, *Application of head-up-display technology to telematics.* Electronics and telecommunications trends, 2008. **23**(1): p. 153-162.
12. Bandara, D., et al., *The application of Augmented Reality Technology in Maritime Operations and Training*, in *International Marine Simulation Forum AGM*. 2019: Johore Bahru, Malaysia.
13. Microsoft. *HoloLens (1st Generation) Hardware* 2020 11th July 2020]; Available from: https://docs.microsoft.com/en-us/hololens/hololens1-hardware.
14. Holder, E. and S.R. Pecota, *Maritime head-up display: a preliminary evaluation.* The Journal of Navigation, 2011. **64**(4): p. 573-594.
15. Hugues, O., J.-M. Cieutat, and P. Guitton. *An experimental augmented reality platform for assisted maritime navigation*. in *Proceedings of the 1st Augmented Human International Conference*. 2010. ACM.
16. Oh, J., S. Park, and O.-S. Kwon, *Advanced Navigation Aids System based on Augmented Reality.* International Journal of e-Navigation and Maritime Economy, 2016. **5**: p. 21-31.
17. ASSOCIATION, I.M.P., *GUIDELINES ON THE DESIGN AND USE OF PORTABLE PILOT*

UNITS, I.M.P. ASSOCIATION, Editor. 2016, International Maritime Pilots' Association (IMPA): HQS Wellington, Temple Stairs, Victoria Embankment, London WC2R 2PN. p. 12.
18. Association, N.M.E. *NMEA 2000® Interface Standard*. [cited 2021 16th February]; Available from: https://www.nmea.org/content/STANDARDS/NMEA_2000.
19. Association, N.M.E. *NMEA 0183 Interface Standard*. [cited 2021 16th February]; Available from: https://www.nmea.org/content/STANDARDS/NMEA_0183_Standard.
20. Gakstatter, E. *What Exactly Is GPS NMEA Data?* February 4, 2015 [cited 2021 16th February]; Available from: https://www.gpsworld.com/what-exactly-is-gps-nmea-data/.
21. Raymond, E.S. *AIVDM/AIVDO protocol decoding*. January 2021 [cited 2021 10th February 2021]; Available from: https://gpsd.gitlab.io/gpsd/AIVDM.html.
22. Bandara, D., et al., *Augmented Reality Lights for Compromised Visibility Navigation.* Journal of Marine Science and Engineering, 2020. **8**(12): p. 1014.

9. AUTHORS BIOGRAPHY

Doupadi Bandara is a PhD researcher at Australian Maritime College, University of Tasmania, Australia. She started her career as an Electrical Engineer in Abans Limited in Sri Lanka from 2007 to 2008. In 2008, she was promoted as a Manager, Electronic Department in Abans Electricals PLC. After 2010, she joined the CINEC Maritime Campus in Sri Lanka as a Lecturer in Department of Electrical and Electronic Engineering. In 2018, I graduated with a Master of Research (Maritime Engineering) from Australian Maritime College, University of Tasmania. Her research interests are in the areas of Augmented Reality (AR), human behaviour, control systems, and image processing.

Michael Woodward is Associate Professor of Marine Engineering at the Australian Maritime Collage. He represents AMC on the Advisory Council of the ITTC, he is the Tasmania Section Representative of RINA and the Pacific region representative of the International Marine Simulation Forum.

Christopher Chin received a B.App.Sci (Hons) in Mathematics from RMIT University in 1997 and also a PhD in Mathematics in 2003. He was appointed a lecturer at the Australian Maritime College (AMC) in 2014. Dr Chin has held several roles over the past few years including First Year Coordinator of the Maritime Engineering degree, Deputy Director (Students & Education) in the National Centre for Maritime Engineering and Hydrodynamics at AMC, Deputy Associate Dean of Learning and Teaching at the College of Sciences and Engineering in the University of Tasmania. Currently, he is a Senior Lecturer at AMC and is the Course Leader for the Maritime Engineering program. Dr Chin's research interests include Engineering Education, renewables, autonomous underwater vehicles and path planning and AR and VR technologies.

Damien Freeman is Facility Manager - Centre for Maritime Simulations, Australian Maritime College, University of Tasmania. He started his working life as an Industrial Toolmaker and Associate Mechanical Engineer. Changing circumstances saw him move into the field of Computer Aided Drafting (CAD) which then evolved into 3D Modelling in the mid 90s. He started work in AMC's Centre for Maritime Simulations in 1997 as a Simulation Modeller. The position of Simulation Modeller is a multidiscipline role involving real time 3D modelling techniques and simulation specific Maritime Geographical Information Systems. With over 24 years of Maritime Simulation experience Damien now leads a team of Kongsberg Digital certified AMC technicians, while providing Simulation technical support for Students, Researchers and Staff.

DATA IMPUTATION OF MISSING VALUES FROM MARINE SYSTEMS SENSOR DATA. EVALUATION, VISUALISATION, AND SENSOR FAILURE DETECTION

C Velasco-Gallego and **I Lazakis**, University of Strathclyde, UK

SUMMARY

To enable Condition-Based maintenance, sensors need to be installed, and thus Internet of Ships (IoS) needs to be implemented. IoS presents several challenges, an example of which is the imputation of missing values. A data assessment imputation framework that is utilised to assess the accuracy of any imputation model is presented. To complement this study, a real-time imputation tool is proposed based on an open-source stack. A case study on a total of 4 machinery systems parameters obtained from sensors installed on a cargo vessel is presented to highlight the implementation of this framework. The multivariate imputation technique is performed by applying Kernel Ridge Regression (KRR). As the explanatory variables may also contain missing values, GA-ARIMA is utilised as the univariate imputation technique. The case study results demonstrate the applicability of the suggested framework in the case of marine machinery systems.

Keywords – data imputation, machine learning, marine machinery systems, condition-based maintenance (CBM), data monitoring, imputation assessment.

NOMENCLATURE

AdaBoost	Adaptive Boosting
ARIMA	Autoregressive Integrated Moving Average
CBM	Condition-Based Maintenance
DAIF	Data Assessment Imputation Framework
DTR	Decision Tree Regression
EWMA	Exponentially Weighted Moving Average
GA	Genetic Algorithm
k-NN	k-Nearest Neighbors
KRR	Kernel Ridge Regression
MAR	Missing at Random
MCAR	Missing Completely at Random
MedAE	Median Absolute Error
MICE	Multiple Imputation by Chained Equations
MNAR	Missing Not at Random
MSE	Mean Squared Error
MSLE	Mean Squared Log Error
NN	Neural Network
OEM	Original Equipment Manufacturers
PLS	Partial Least Squares
RBF	Radial Basis Function
RMSE	Root Mean Square Error
STL	Seasonal and Trend decomposition using Loess
SVR	Support Vector Regression
VAR	Vector autoregression

1. INTRODUCTION

In 2019 alone a total of 2904 casualties and 49 fatalities occurred in relation to maritime transportation. Of all the causes of accidents to ships, 14% refers to damage to ship equipment [1]. Between 75% and 96% of marine accidents are the result of human error [2]. In consideration of climate change concerns, CO_2 emissions are expected to increase between 50% and 250% by 2050 if no measure is implemented [3]. Hence, there is no doubt whatsoever about the improvement threshold that can be implemented within the maritime industry.

In relation to maintenance, the maritime industry is currently considering state-of-the-art maintenance and inspection processes, an example of which is Condition-Based Maintenance (CBM). This is a strategy hinged on the condition monitoring of the asssets [4][5][6][7][8]. To enable this strategy, sensors need to be installed, and thus Internet of Ships (IoS) needs to be implemented. IoS presents several challenges, an example of which is the imputation of missing values [9][10][11].

Data imputation is a compelling pre-processing step, the aim of which is the estimation of identified missing values to avoid under-utilisation of data. Hence, if missing values are not tackled, the results obtained from applying data analysis may be unreliable and inaccurate, which could lead to bias in further steps due to the utilisation of poor models implemented in decision-making processes [12].

Although there are more than 10,000 publications about data imputation in Scopus database, only two publications refer to the maritime industry. This indicates a lack of analysis and formalisation of data imputation in this industrial sector. For this reason, a Data Assessment Imputation Framework (DAIF) is developed to assess the accuracy of any imputation model. Thus, the selection of the imputation approach is not biased by human decisions and its selection is purely objective based on the characteristics and contexts of the data.

In addition, a real-time imputation tool is presented based on an open-source stack, so that any organisation can implement this framework to monitor sensor data, and thus prevent sensor failure. The application of both the DAIF and the real-time imputation tool is highlighted through a case study, in which missing values from the main engine power parameter are imputed.

The following paragraphs are structured as follows. Section 2 presents data imputation studies performed within the maritime industry. Section 3 describes the proposed methodology. Section 4 reflects on the result obtained after implementing the proposed methodology through a case study. Lastly, in Section 5 the conclusions are presented.

2. LITERATURE REVIEW

A total of two articles were identified from Clarivate Analytics Web of Science and Scopus. [14] presented a hybrid imputation method by combining *k*-Nearest Neighbors (*k*-NN) and Multiple Imputation by Chained Equations (MICE) models. A total of three metrics (Absolute Percentage Error (APE), Mean Absolute Percentage Error (MAPE), and the standard deviation of the error) were estimated to compare the proposed framework with *k*-NN and MICE algorithms. A case study based on the imputation of missing values from time-series data collected from a total of 8 sensors coupled to the turbocharger and to the main engine of a chemical tanker was implemented to demonstrate that the proposed methodology outperforms *k*-NN and MICE methods.

[13] performed a comparative study that examined a total of 20 widely implemented machine learning and time series forecasting algorithms (mean imputation, Seasonal and Trend using Loess (STL) decomposition, exponential smoothing, Autoregressive Integrated Moving Average (ARIMA) models, linear regression models (Partial Least Squares (PLS) regression, Least Absolute Shrinkage and Selection Operator (LASSO) regression, Ridge regression, and ElasticNet regression, *k*-NN, Support Vector Regression (SVR) with linear and Radial Basis Function (RBF) kernels, Neural Networks (NNs) with 1, 2, and 3 hidden layers, Vector autoregression (VAR), Decision Tree Regression (DTR), and ensemble methods (bagged trees with SVR, bagged trees with *k*-NN, random forests, and Adaptive Boosting (AdaBoost)). A case study was also implemented on a total of 7 machinery parameters obtained from sensors installed on a cargo vessel to assess their performance. It was concluded that ARIMA outperformed the remaining models.

Although both articles presented new methodologies to impute missing values from sensor data of marine machinery, there are yet several challenges to be tackled within the maritime industry. An example of which is the development of a generic framework to assess the performance of any imputation model, novelty that is introduced in this paper. A list of the novelties presented in this paper in relation to data imputation in the maritime industry is described hereunder.

- The development of a data assessment imputation framework to evaluate the accuracy of any imputation model.
- The implementation of Exponentially Weighted Moving Average (EWMA) as a denoising method in the pre-processing step to enhance the performance of the data imputation model.
- The application of the Kernel Ridge Regression (KRR) as a multivariate data imputation model.
- The employment of GA-ARIMA as a univariate data imputation model.
- The proposal of a real-time imputation tool based on an open-source stack.

3. METHODOLOGY

Having explored the novelties presented in this conference paper, this section presents the proposed framework, which is graphically represented in Figure 1.

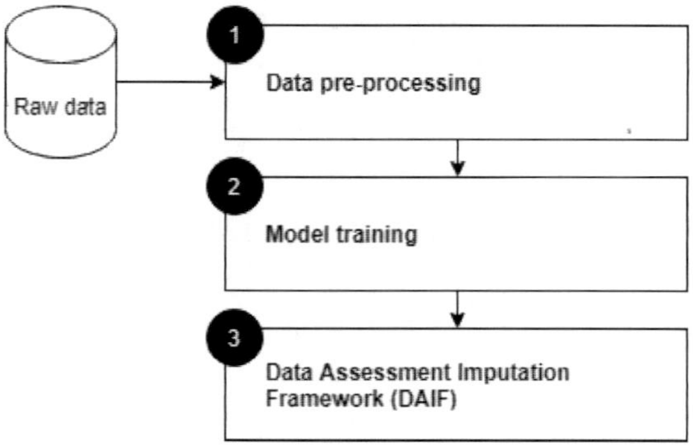

Figure 1. Graphical representation of the proposed methodology.

3.1 DATA PRE-PROCESSING

Data pre-processing is essential to be implemented prior to model training due to the characteristics of the data, as the raw data may not only include steady operational states but also both manoeuvring and transient states of machinery, which need to be excluded from the analysis. To define proper steady operational states Original Equipment Manufacturers (OEMs) of the systems being analysed are consulted.

Subsequently, standardisation is applied so that features can contribute equally when distance-based algorithms are implemented. Correlation analysis is also performed by the estimation of both the Pearson's correlation coefficient and Spearman's rank correlation coefficient to identify linear and non-linear relationships between features. Feature extraction has been performed prior to all these steps and only four parameters of the main engine system of a cargo vessel are included in this study (main engine power, main engine rotational speed, main engine fuel flow rate, and scavenging air pressure of the scavenge air receiver). It is assumed that all occurrences of the sample refer to normal operational conditions.

In addition, as time series data are susceptible to containing high noise, Exponential Weighted Moving Average (EWMA) is implemented as a denoising technique.

3.2 MODEL TRAINING

The model implemented to impute missing values in this study is the Kernel Ridge Regression (KRR), which combines Ridge regression with the kernel tricks. Ridge regression is a biased model, as it adds a penalty to regularise the parameter estimates, and thus apply a trade-off between the bias and the variance to reduce the Mean Squared Error (MSE). Ridge regression considers the addition of a second-order penalty on the parameter estimates. Among all the kernel functions that encompass linear and non-linear functions of the predictors, the Radial Basis Function (RBF) is considered in this study. Hence, two main hyperparameters need to be selected optimally (regularisation strength, α, and gamma parameter for the RBF kernel, γ). To that end, a Genetic Algorithm (GA) is applied, which is a variety of evolutionary algorithms inspired by the process of natural selection.

The selection of these two hyperparameters is essential as an inadequate selection of them may yield either under-fitting or over-fitting of the KRR model, and thus lead to inaccurate imputations. Therefore, time series cross-validation is implemented to select the hyperparameters that best conduct the learning process.

3.3 DATA ASSESSMENT IMPUTATION FRAMEWORK (DAIF)

In general, a total of three different missingness mechanisms have been identified to encompass underlying causes of missing data.

The first mechanism, Missing Completely at Random (MCAR), involves those situations in which the missingness is independent of the data. An example of which is a random failure produced in the fuel flowmeter [14]. To simulate this situation, the following procedure is reproduced:

- n samples with different missing ratios (r_1, r_2, ..., r_m) are generated. Each sample contain values missing completely at random.
- Univariate imputation is performed to impute missing values from predictors' instances. The method applied is GA-ARIMA. GA-ARIMA is a variation of Autoregressive Integrated Moving Average (ARIMA) models, in which GA is utilised to estimate the coefficients.
- Multivariate imputation is performed as indicated in the preceding section to impute missing values from the response variable.

The second mechanism is Missing at Random (RAM), in which the missingness hinges on another feature. For instance, if a component of a main engine fails, the operating condition of dependent components may be altered. This case is simulated by applying the strategy expressed hereunder.

- n large gaps of m size are genereted at random. As a considerable amount of successive occurrences are missing for each feature, it is highly probable that instances contain more than one feature, the occurrences of which are missing.
- Analogous to the MCAR mechanism, univariate imputation is implemented, and thus missing values from predictors' instances are imputed. The technique applied to perform the univariate imputation is GA-ARIMA.
- Multivariate imputation is performed as indicated in the preceding section to impute missing values from the response variable.

The third and final mechanism is established as Missing Not at Random (MNAR) and refers to those scenarios in which the missingness is related to the feature itself. An example of which is the failure of a component, which is unable to record operating data as it is malfunctioning. However, as indicated in the preceding sections, some assumptions have been made to conduct this study, an example of which is the absence of fault data. Thus, this mechanism is out of the scope of this paper.

To evaluate the accuracy of the imputations performed in this section of the methodology, a total of five metrics can be estimated (Root Mean Square Error (RMSE), MSE, max error, Median Absolute Error (MedAE), Mean Squared Log Error (MSLE)). Once the selected metric has been estimated for all the imputation scenarios, the results are pooled to obtain a unique value that summarises the imputation performance of the model being tested.

4. RESULTS

To highlight the implementation of the methodology described above, a case study is presented. To that end, data have been collected from a DMD-MAN B&W 6S50MC-C main propulsion engine of a cargo vessel. Specifically, the main engine power, the main engine rotational speed, the main engine fuel flow rate, and the scavenging air pressure of the scavenge air receiver parameters have been considered for this case study. Time series plots of all four parameters can be observed in Figure 2.

The samples contain a total of 1,200 records, which refer to the steady operational states of machinery. As observed in Figure 2, a total of four steady states are identified in the samples. For instance, if the main engine power is considered, there is a large steady state that initiates at the first instant and persists around 200 minutes where the values are stabilised around 4,200 kW. Subsequently, there is an adjustment where the main engine power decreases to approximately 3,800 kW. This state remains for roughly 700 minutes, and then suddenly an abrupt increase is perceived. This is when the maximum value of the time series is achieved, which presents a value greater than 4,500 kW. This state remains for 200 minutes. Finally, the last abrupt change occurs. This is when the minimum value of the time series is achieved, which presents a value lower than 3,800 kW. Analogous patterns are observed in the remaining parameters being analysed. All the adjustments observed are applied due to the contractual agreements between the charterer and the shipowner in relation to the vessel speed and the fuel oil consumption per day.

Summary statistics of each time series can also be consulted by analysing the descriptive statistics in Table 1. Correlation analysis has also been performed and the resulting Pearson's and Spearman's correlation matrices can be consulted in Tables 2 and 3, respectively. As indicated, all the parameters are linearly correlated.

To finalise the data pre-processing step, denoising has been applied by implementing EWMA to remove the high noise that time-series data contain. The results of which can also be perceived in Figure 2.

The second step of the methodology is model training, in which KRR model is trained to be used as a multivariate imputation model. 5-fold time-series cross-validation technique is applied to prevent either under-fitting or over-fitting. As two main hyperparameters need to be estimated (α and γ), an optimisation algorithm needs to be implemented. For this study, GA is applied. The sets considered to perform GA are expressed hereunder.

$\alpha \in [10^{-10}, \ldots, 1]$
$\gamma \in [10^{-10}, \ldots, 10^{-3}]$

Due to the extent limitations of the present conference paper, only the main engine power is considered for assessing the imputation of missing values, and thus the three remaining parameters (main engine rotational speed, main engine fuel flow rate, and scavenging air pressure of the scavenge air receiver) are only utilised as explanatory variables. However, the presented framework can be used to assess the imputation performance for any parameter.

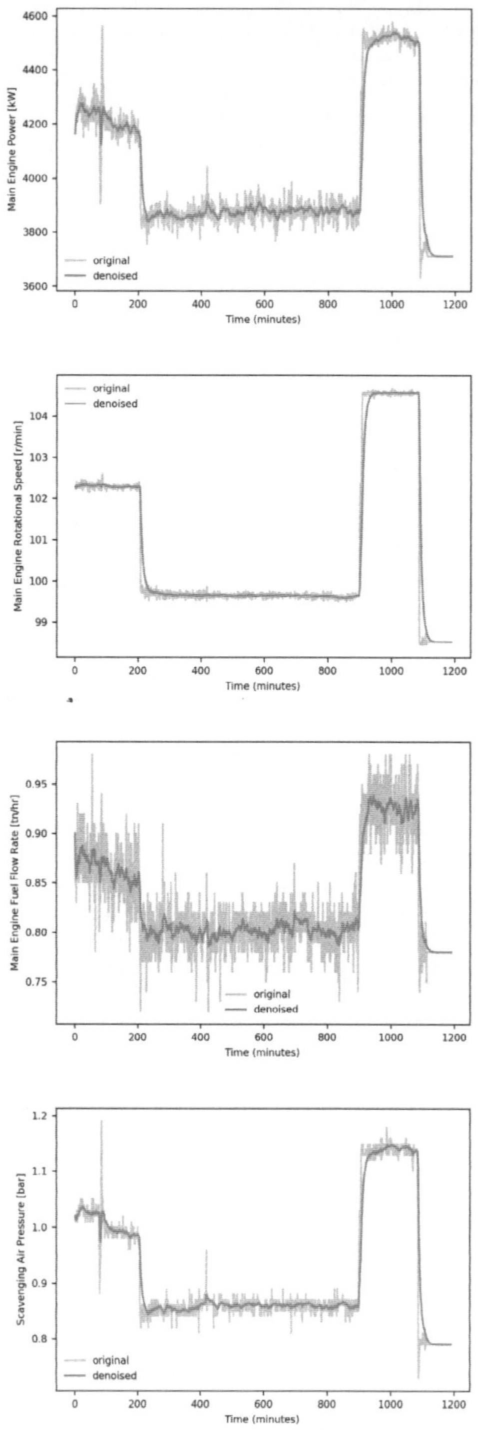

Figure 2. Time series plot of (a) the main engine power, (b) the main engine rotational speed, (c) the main engine fuel flow rate, and (d) the scavenging air pressure of the scavenge air receiver.

To assess the imputation performance of KRR, the data assessment imputation framework is implemented. The ratios considered to generate the MCAR samples are 0.05, 0.15, 0.20, 0.25, and 0.3. In relation to the MAR samples, a total of 5 large gaps, the sizes of which are between 50 and 150 occurrences, are generated at random for each analysed sample. These values are selected based on the characteristics of missingness and contexts perceived in the data collected, as well as suggestions made by other academics who have analysed marine machinery data.

To compare the imputation performance of KRR, GA-ARIMA is also implemented. The criteria utilised to select this model for the comparative study is due to its implementation as a univariate imputation technique to impute missing values from predictors' instances. In addition, ARIMA models presented the best results in the case study performed in [13]. For each model two variations are considered: 1) denoising is implemented in the pre-processing step by the implementation of EWMA, and 2) no denoising is implemented.

MCAR results for each fold can be observed in Table 4. Table 5 presents both large gaps and final results of each implemented model.

As it can be perceived, and as deduced in [13], ARIMA models outperform multivariate imputation techniques when MCAR context is considered. For all MCAR scenarios, ARIMA achieved a RMSE no greater than 30 kW, with the exception of the cases introduced in fold 5, in which the maximum error (406.08 kW, ratio 0.20) is achieved. This increment can be perceived for all four approaches, which indicates that these substantive alterations are related with the characteristics of the time series. As identified in Figure 2, there are a total of four steady states, the last one being the shortest one. This steady state is not considered in the training set of any of the models, and thus the predictions cannot be constructed adequately. This is an indicator of how important it is to prevent failures that lead to the collection of either incorrect or missing values, as this may yield to biased imputations if analogous preceding instances have not been perceived.

Another consideration that can be distinguished is the enhancement of the imputation when denoising is applied. For instance, when this pre-processing step is implemented prior to the imputation of missing values by the utilisation of GA-ARIMA a percentage of improvement of approximately 50% can be observed. This indicates that denoising may need to be applied when a model sensitive to noise is being performed and the time-series data being analysed contain high noise.

In relation to MAR results, KRR leads to better results than GA-ARIMA, as GA-ARIMA cannot predict unexpected events. Thus, GA-ARIMA is unable to detect when an adjustment is introduced due to either the contractual agreements between the charterer and the shipowner or weather conditions. For this reason, multivariate imputation techniques are recommended when the instances with missing values are perceived and when unexpected events occur. This also highlights the importance of preventing failures that lead to the collection of either incorrect or missing values, as biased imputations may be obtained if the unexpected events are not recorded.

Therefore, as indicated in the preceding paragraphs, there is no unique model that outperforms the remaining imputation techniques for all characteristics and contexts. Hence, the implementation of comparative studies and assessment frameworks to determine the best model to impute missing values based on the scenario presented is of paramount importance, as under-utilisation of data can then be prevented. Furthermore, the frameworks applied can be part of a holistic predictive framework, in which diagnosis and prognosis can be performed to assess the current and future health of machinery to assist instant decision-making processes. Thus, maintenance and inspection tasks, crew management, and spare parts stocks can be optimised.

To make profit of all this potential, the proposed frameworks are combined with data warehouse and data dashboard solutions. The implementation of data dashboards is considered one of the three critical aspects (data warehouse, data dashboard, and training development) that need to be implemented to advance a company towards data maturity. Data warehouse is the core component of business intelligence, utilised for data analysis and reporting. Data dashboard is an information management tool that supports a specific insight based on the extraction, analysis, measurement, and monitoring of data. As data is being democratised, there is also a necessity to adapt the culture of the company to orient decision-making strategies toward data-driven approaches. Thus, the development of new skills in the personnel is of paramount importance to guarantee the success of this transition.

Accordingly, to finalise the results section, an example of a data dashboard developed based on the open-source stack is presented in Figure 4. As observed, instances of the four parameters analysed in this study are graphically represented by the implementation of Grafana software (https://grafana.com). These visualisations can be utilised by the personnel to identify missing values, and thus implement inspection activities if needed. Alarms can also be set so that a message is sent to the personnel when a certain amount of missing values is achieved per hour. Hence, sensor failure can be detected at an early stage or even prevented, guaranteeing the quality of data required to perform data-driven decision-making strategies.

Table 1. Descriptive statistics of the monitored features.

	Power [kW]	Speed [rpm]	Fuel Flow Rate [tn/hr]	Scav. Air Press. [bar]
Count	1300	1300	1300	1300
Mean	4415.28	104.19	0.87	1.09
Std.	41.94	0.10	0.02	0.03
Min.	4313.49	103.92	0.81	1.02
25%	4387.015	104.12	0.86	1.07
50%	4408.425	104.19	0.87	1.09
75%	4433.1675	104.27	0.88	1.11
Max.	4566.69	104.51	0.93	1.19

Table 2. Pearson's correlation matrix.

	Power [kW]	Speed [rpm]	Fuel Flow Rate [tn/hr]	Scav. Air Press. [bar]
Power [kW]	-	0.99	0.90	0.99
Speed [rpm]	0.99	-	0.90	0.99
Fuel Flow Rate [tn/hr]	0.90	0.90	-	0.91
Scav. Air Press. [bar]	0.99	0.99	0.91	-

Table 3. Spearman's correlation matrix.

	Power [kW]	Speed [rpm]	Fuel Flow Rate [tn/hr]	Scav. Air Press. [bar]
Power [kW]	-	0.80	0.76	0.90
Speed [rpm]	0.80	-	0.74	0.82
Fuel Flow Rate [tn/hr]	0.76	0.74	-	0.78
Scav. Air Press. [bar]	0.90	0.82	0.78	-

Table 4. a) MCAR results fold 1.

	0.05	0.15	0.20	0.25	0.30
GA-ARIMA (EWMA)	21.17	13.69	16.55	20.01	20.19
GA-ARIMA	38.29	41.15	41.18	38.35	42.34
KRR (EWMA)	245.06	320.97	291.54	309.98	290.15
KRR	272.49	343.85	339.18	320.08	326.70

Table 4. b) MCAR results fold 2.

	0.05	0.15	0.20	0.25	0.30
GA-ARIMA (EWMA)	12.127	12.070	11.732	14.015	14.825
GA-ARIMA	45.468	48.617	53.193	53.739	54.790
KRR (EWMA)	365.439	372.451	368.937	373.311	373.738
KRR	393.447	386.104	393.150	394.653	384.923

Table 4. c) MCAR results fold 3.

	0.05	0.15	0.20	0.25	0.30
GA-ARIMA (EWMA)	21.17	16.97	16.85	26.43	23.91
GA-ARIMA	47.63	45.04	55.38	45.90	46.48
KRR (EWMA)	365.08	365.27	368.05	366.35	365.03
KRR	340.91	326.56	342.13	321.13	324.75

Table 4. d) MCAR results fold 4.

	0.05	0.15	0.20	0.25	0.30
GA-ARIMA (EWMA)	16.36	22.23	23.53	25.64	23.00
GA-ARIMA	41.13	42.29	114.52	144.51	80.47
KRR (EWMA)	332.94	329.03	329.55	328.72	325.09
KRR	323.93	317.42	317.78	328.38	322.54

Table 4. e) MCAR results fold 5.

	0.05	**0.15**	**0.20**	**0.25**	**0.30**
GA-ARIMA (EWMA)	347.88	370.19	406.08	402.20	389.98
GA-ARIMA	575.34	608.99	568.87	567.17	603.60
KRR (EWMA)	295.18	289.66	297.84	295.90	322.83
KRR	391.58	409.42	422.26	399.49	427.91

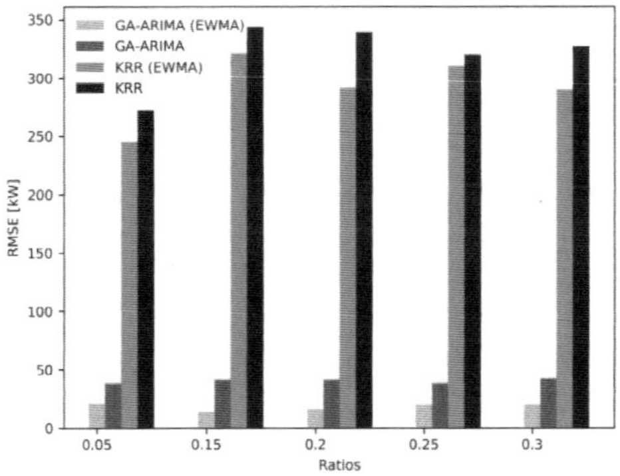

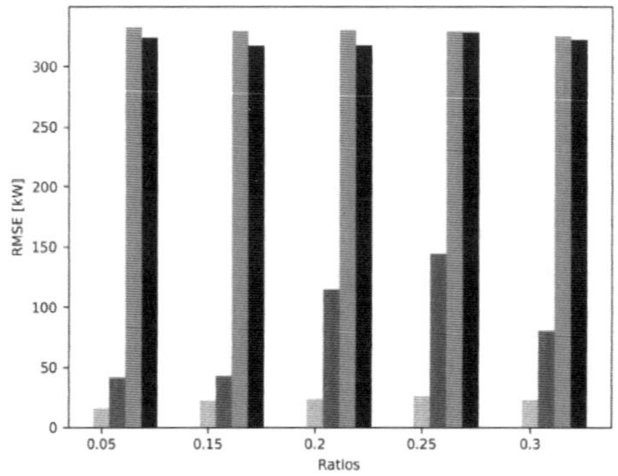

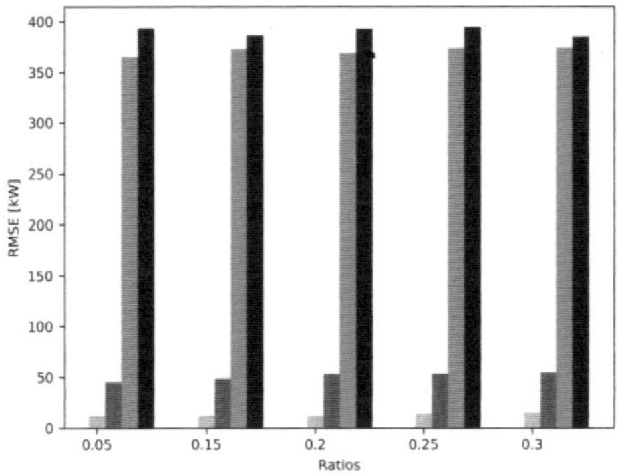

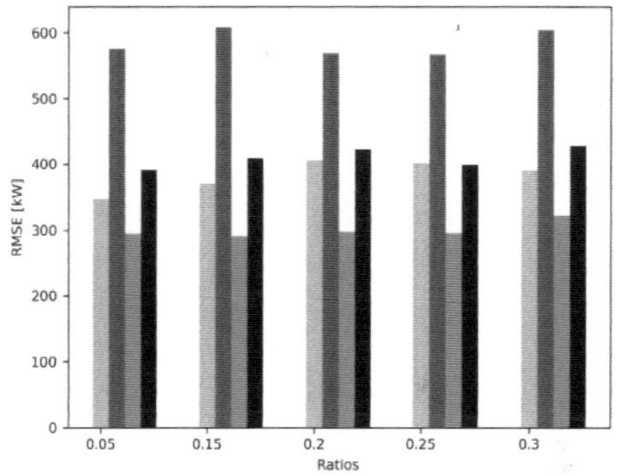

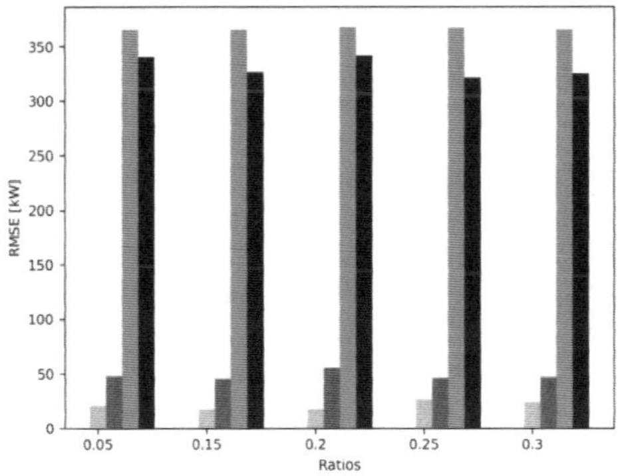

Figure 3. MCAR results plot of (a) fold 1, (b) fold 2, (c) fold 3, (d) fold 4, and (e) fold 5.

Table 5. DAIF results.

	MCAR	MAR	Final
GA-ARIMA (EWMA)	91.67	257.99	174.83
GA-ARIMA	161.634	279.41	220.52
KRR (EWMA)	331.524	234.24	282.88
KRR	354.83	201.26	278.05

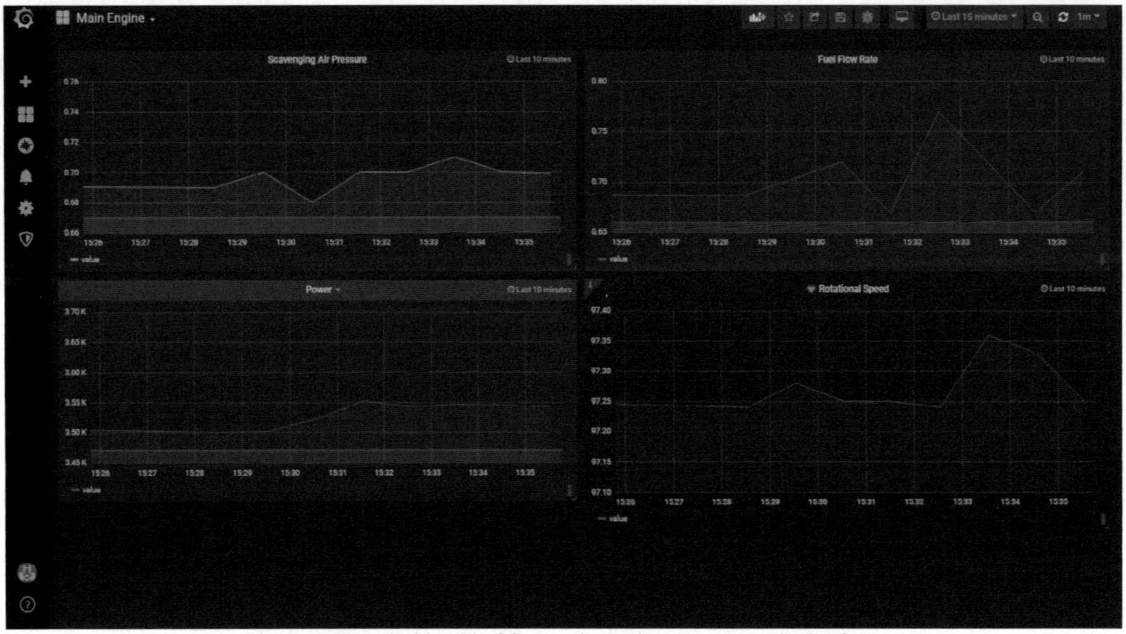

Figure 4. Data dashboard of four main engine parameters in Grafana.

5. CONCLUSIONS

The maritime industry is currently considering state-of-the-art maintenance and inspection processes, an example of which is CBM. To enable this maintenance strategy, sensors need to be installed, and thus IoS needs to be implemented. IoS presents several challenges, including the imputation of missing values.

As previously stated, while there are more than 10,000 publications about data imputation in Scopus database, only two of these refer to the maritime industry. This indicates a lack of analysis and formalisation of data imputation in this industrial sector. To contribute to the development of a data imputation framework in the maritime industry, a total of five novelties were proposed: 1) the development of a data assessment imputation framework to evaluate the accuracy of any imputation model, 2) the implementation of EWMA as a denoising method in the pre-processing step to enhance the performance of the data imputation model, 3) the application of KRR as a multivariate imputation model, 4) the employment of GA-ARIMA as a univariate data imputation model, and 5) the proposal of a real-time imputation tool based on an open-source stack.

In relation to these novelties, the main conclusions are expressed hereunder.

- There is not a unique model that outperforms the remaining imputation techniques for all possible characteristics and contexts described in the maritime industry. Hence, the implementation of comparative studies and assessment frameworks is of paramount importance.
- KRR leads to better results when large gaps of missing values need to be imputed, as GA-ARIMA cannot predict unexpected events.
- GA-ARIMA outperforms KRR when MCAR context is considered.
- The enhancement in the performance that can be produced when denoising is applied to time-series data that contain high noise and when a model sensitive to noise is implemented.
- The importance of preventing failures that lead to the collection of either incorrect or missing values, as this may yield to biased imputations if analogous preceding intances have not been perceived and unexpected events are not recorded.
- The influence of an effective data dashboard on the prevention of sensor failure.

Although some novelties are presented in this paper, further research needs to be addressed due to the importance of this pre-processing step. For this reason, the authors of this paper suggest the following work guidelines:

- A comprehensive comparative methodology of deep learning models as imputation techniques.
- The development of a software that provides data imputation tool to be included in a holistic predictive framework to assist real-time data-driven decision-making strategies.
- The analysis of optimisation algorithms for tuning hyperparameters.
- The study of denoising alogirithms.

6. REFERENCES

1. EUROPEAN MARITIME SAFETY AGENCY (EMSA), 'Preliminary annual overview of marine casualties and incidents', 2020.

2. ELLEFSEN, A. L. ÆSØY, V., USHAKOV S., ZHANG H., 'A Comprehensive Survey of Prognostics and Health Management Based on Deep Learning for Autonomous Ships', *IEEE Transactions on reliability*, 2019.

3. INTERNATIONAL MARITIME ORGANIZATION (IMO), 'Greenhouse Gas Emissions', 2020.

4. LAZAKIS I., DIKIS K., MICHALA A.L., THEOTOKATOS G., 'Advanced ship systems condition monitoring for enhanced inspections, maintenance and decision making in ship operations', *Transportation Research Procedia*, 2016.

5. LAZAKIS I., GKEREKOS C., THEOTOKATOS G., 'Investigating an SVM-driven, one-class approach to estimating ship systems condition', *Ships and Offshore Structures*, 2018.

6. LAZAKIS I., RAPTODIMOS Y., VARELAS T., 'Predicting ship machinery system condition through analytical reliability tools and artificial neural networks', *Ocean Engineering*, 2018.

7. RAPTODIMOS Y., LAZAKIS I., 'Using artificial neural network self-organising map for data clustering of marine engine condition monitoring applications', *Ships and Offshore Structures*, 2018.

8. RAPTODIMOS Y., LAZAKIS I., 'Application of NARX neural network for predicting marine engine performance parameters', *Ships and Offshore Structures*, 2019.

9. BALAKRISHNAN S.M., SANGAIA A.K., 'Chapter 6 – aspect oriented modelling of missing data imputation for Internet of Things (IoT) based healthcare infrastructure', *Intelligent Data-Centric*, 2018.

10. IZONIN I., KRYVINSKA N., TKACHENKO R., ZUB K., 'An approach towards missing data recovery within IoT smart systems', *Procedia Computer Science*, 2019.

11. NOOR N.M., ABDULLAH M.M.A.B, YAHAYA A.S., RAMLI N.A., 'Comparison of linear interpolation method and mean method to replace missing values in environmental data set', *Mater. Sci.*, 2014.

12. FEKADE B., MAKSYMYUK T., KYRYK M., JO M., 'Probabilistic recovery of incomplete sensed data in IoT', *IEEE Internet of Things Journal*, 2018.

13. VELASCO-GALLEGO, C., LAZAKIS, I., 'Real-time data-driven missing data imputation for short-term sensor data of marine systems. A comparative study', *Ocean Engineering*, 2020.

14. CHELIOTIS M., GKEREKOS, C., LAZAKIS, I., THEOTOKATOS, G., 'A novel condition and performance hybrid imputation method for energy efficient operations of marine systems', *Ocean Engineering*, 2019.

7. AUTHORS BIOGRAPHY

Mr. Christian Velasco-Gallego is a PhD student in the Department of Naval Architecture, Ocean, and Marine Engineering at the University of Strathclyde. He has experience from working in multinationals as a technical consultant and supply chain specialist. His research interests include predictive maintenance and supply chain management.

Dr. Iraklis Lazakis, Reader, CEng, MBA, FRINA, MSNAME, FHEA, PhD is a Reader at the Department of Naval Architecture, Ocean, and Marine Engineering at the University of Strathclyde. He has extensive academic and industry experience working in UK and EU research and marine industry projects. His research interests include ship systems condition monitoring, systems maintenance and reliability, ship operations, risk analysis, shipyard manufacturing and productivity, and asset management.

INNOVATION IN SOFTWARE. THE COMPLEXITY OF THE CHANGES THAT INDUSTRY 4.0 IS DEMANDING

R Pérez Fernández, SENER, Spain

SUMMARY

Industry 4.0 has favoured the expansion of many technologies whose application boundaries are very diffuse. Although each technology can have very specific applications, when it is wanted to apply to *Computer Aided Design*, *Manufacturing & Engineering* (*CAD*/CAM/CAE) System, they have to be implemented as a whole.
The first *CAD* tools were developed during the middle part of the 20th century, and since then they are in the beginning of the design, but it manages many data that must be taken in advance for the further stages of the lifecycle of the product. *Augmented Reality*, *Virtual Reality* and *Mixed Reality* are closely related to the *Digital Twin* and interlaced with the *Big Data*, which are generated by *CAD* tools and all surrounding solutions, which applies some cloud/edge/fog computing to these data in a merged technology between finite state machines and *Artificial Intelligence* (AI) cognitive processes.
To perform in an agile manner all these integrations requires a network which support different connection ways to add specific devices, i.e. *Internet of Things* (*IoT*), which can access to the data, creating and modifying them, in a different layer which affects to the basic information layer created by the *CAD* System in the shipyard.
This network should be secured but at the same time open to allow distributed work, which must be tracked in such a way that records any modification of every working step done in an open, transparent, trusted and non-modifiable working method for all stakeholders involved in process, like: shipyard, engineering offices, classification society and ship owner. Results of the design should be easy integrated with future building ways like *3D* printing, generating printing orders directly from the *CAD* model.
All the above technologies are reviewed in this paper, as well as revisit the application of them in every shipbuilding phase, from design to manufacturing. An integrated *Industry 4.0 CAD* system should also be involved in operation and maintenance phases, i.e. it must cover all the vessel lifecycle end to end, from design to decommission.

NOMENCLATURE

CAD	*Computer Aided Design*
IoT	*Internet of Things*
IoS	*Internet of Ships*
PLM	*Product Lifecycle Management*

1. HISTORICAL BACKGROUND AS REGARDS CAD SYSTEMS

During the Third Industrial Revolution, which started during the 1950´s, the *Computer Aided Design* (CAD) systems started to be developed in order to facilitate the generation of hull forms.

The basis of the first *CAD* tools were the mathematical formulation of hull forms, which was used for both form fairing and form generation.

For selecting the function representing the waterlines for both the forebody and afterbody, the following basic requirements were established:

- Continuity of the function and its derivatives, at least up to the second order.
- Ability to represent the parallel part both in the fore and after bodies.
- Easy control of number and position of inflection points.
- Low number of parameters.
- Clear geometrical or physical meaning for each parameter.
- Suitability for further mathematical treatment to solve basic hydrodynamic problems, particularly as refers to theoretical calculation of wave-making resistance.
- Versatility for accurately representing a wide range of waterlines from a selection of many existing ship forms from various origins.

Although polynomic functions were considered in the beginning, they were soon discarded on account of the high number of parameters and difficulty of con trolling inflection points. Many other types of algebraic functions were then tested without success.

Finally, a number of non-algebraic functions were investigated, one of which was found to give fully satisfactory results.

Although, for obvious reasons, full details cannot be given here, the basic peculiarities of the mathematical formulation can be seen in *figure 1*, which represents a waterline in the fore or after body.

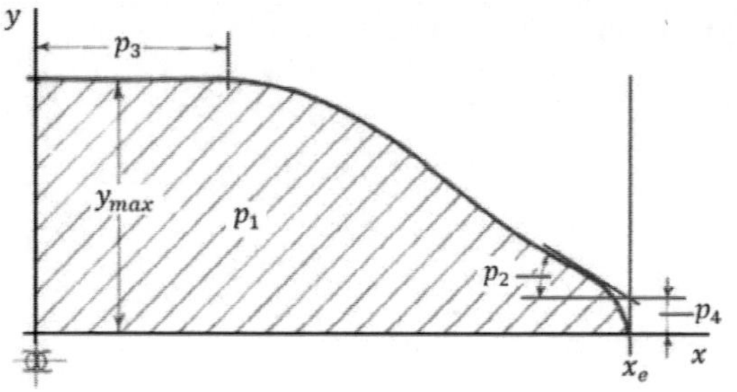

Figure 1: Representation of a waterline in the fore or after body

x is measured from the midship section to the tore or after end of the ship.

y is measured from amidships to the port or starboard side. x_e e and y_{max} are the maximum x and y coordinates of the waterlines.

Then, using non-dimensional coordinates:

$$x = \frac{x}{L_{PP}/_2} \tag{1}$$

$$y = \frac{y}{y_{max}} \tag{2}$$

the non-dimensional half-breadth for any point of the waterline can be expressed by an explicit non-algebraic function of the type:

$$y = y\,(p_1, p_2, p_3, p_4, x, x_e) \tag{3}$$

Where:

p_1 represents the coefficient of area of the waterline.

p_2 represents the non-dimensional angle of entrance (for the forebody) or run (for the afterbody).

p_3 represents the non-dimensional length of the parallel part of the waterline.

p_4 represents the non-dimensional half-width of the prolonged waterline for the abscissa x_e.

The above function satisfies the requirements established as:

- All derivatives are continuous.
- The number of inflection points is zero, one or two, depending of the values of the parameters. There are only
- Four parameters, each with a clear geometrical meaning. For the waterlines investigated, the mean square error of $(y\text{-}y_{actual})$ ranges from 1 x 10^5 to 4 x 10^{-3}. The highest deviations occur for ships with rather poor hydrodynamic characteristics, such as the fullest form in Series 60.

As already mentioned, the first area with which *CAD* systems was concerned was that of hull forms. It is well known that the possibility of generating hull forms defined by mathematical equations has attracted the attention of the profession for many years. At the beginning of last century, Admiral D.W. Taylor used algebraic functions to generate waterlines and transverse sections of ships.

In spite of the great merit of Taylor's work, his methods were never widely used, probably due to their lack of generality, and to the cumbersome calculations involved.

It seemed surprising, when the availability of fast electronic computers removed the last mentioned limitation to the use of methods developed along Taylor's lines, very little attention was being paid to improve such form generation methods, as far as generality was concerned. Instead, a lot of effort was being devoted to develop numerical hull fairing methods, which could replace the classical manual ones.

After seen how the *CAD* Industry started, it will be seen the technologies which will be implemented in future *CAD* systems.

2. NEW TECHNOLOGIES AS PART OF THE INDUSTRY 4.0

The *Fourth Industrial Revolution* takes the automation of manufacturing processes to a new level by introducing customized and flexible mass production technologies.

This transformation implicates a huge collection of interrelated and joined technologies, which can be analysed independently, but must be applied as a complete integrated implementation in each industrial field, moreover, in the *CAD* development industry. Each technology expose a set of boundaries which cannot be differentiate evidently from its neighbouring technology.

Virtual Reality, *Augmented Reality* and *Mixed Reality* are closely associated to the *Digital Twin* and interconnected with the *Big Data*, which is produced by the *CAD* tools and all surrounding solutions, which applies some *cloud/edge/fog* computing to this data in a merged technology between finite state machines and *Artificial Intelligence* cognitive processes.

To perform in an agile manner all these computing, it requires a network which support different connection ways to add special devices, i.e. *Internet of Things* (*IoT*), which can access to the data, creating and modifying it, in a different layer which affects to the basic information layer created by the *CAD* System in the shipyard.

This network should be secured, but open to allow distributed work, which must be step controlled in a manner that records any modification of each working step done in an open, transparent, trusted and non-modifiable working method for all actors involved in process, like: shipyard, engineering offices, classification society and ship owner.

Results of the design should be easy integrated with future building ways like *3D printing*, generating printing orders directly from the *CAD* model or from the *Product Lifecycle Management* (PLM).

Shipbuilding engineering phases involve design and production, but an integrated *Industry 4.0 CAD* System should also be involved in operation and maintenance.

3. INTERNET OF SHIPS. THE IoT APPLIED TO THE SHIPBUILDING ENVIRONMENT

The success of *IoT* will only be possible if the initiatives have a clear component oriented to the business. It is necessary to identify which of the initiatives provide a clear value to the business, but taking into account that this value can be in various terms and not all directly economic. The clearest ones have to do with the optimization of energy costs, fuel consumption, choice of routes, safety at sea, but also in the work itself inside the ship (*Muñoz and Pérez, 2017*).

One of the conclusions of all the analysis about the consequences of the *IoT* is that objects must be re-thought from the early design.

The fact is that ships are subject to harsh environmental conditions makes that any technological advance to be applied inside a ship have to take into account from the very beginning. Currently all the designs in the modern society are made by information technology tools, computer aided designs, supported with adequate databases and with lifecycle management of the information. To address the challenge of the *IoT* for ships, it is necessary a new concept, the *Internet of Ships* (IoS).

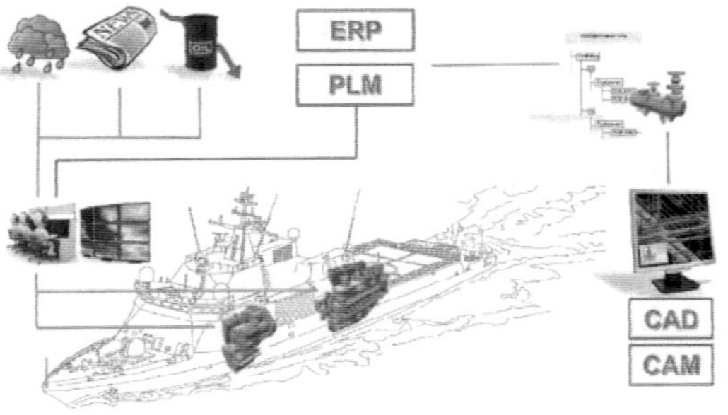

Figure 2: Ideal representation of a 3D model with access to the different ship design disciplines

In this context, the question is if the naval sector is ready for this revolution. Is it possible that this traditional and conservative sector moves into this technology? There is already evidence that the shipbuilding industry is no stranger to these developments and is already connected to the Internet some components of ships, as it is exposed on *figure 2*.

Ship owners can monitor the vessel's status in real time and apply analytics to current and historical data to make decisions that enable them to run more efficiently, saving time and fuel.

Nowadays *CAD* and *PLM* solutions can be used in a pocket tools, making it the indispensable ally in this new technological revolution.

Shipbuilding process generates a lot of information and data, which a priori makes it seem impossible to have all this data in real time, but the new processors, simpler and smaller, with a good connection to the Internet, make it possible.

The data management is, however, only one side of the coin of the *IoS*. Energy efficiency is a fundamental aspect also in new devices that connect to the network.

But *IoS* not only covers the stages of design or production of the boat. Once the sensors are in the components whose information want to monitor, we will be able to obtain information throughout the life of the ship.

IoS is presented as a solution capable of detecting when a component on a boat is close to fail and must be replace, when we take the boat to repair when we have to paint again, when corrosion has reached a certain limit ... and all this from our pocket tool and early enough to avoid late or unforeseen performances. *IoS* reaches this sector to ensure profitable production, or safe, efficient and sustainable process for all types of fishing vessels, tugboats, tankers, charges, ferries, dredgers and oceanographic ...

4. SIMULATION TOOLS

During the engineering phase, simulation tools and model tests are intensively used:

- **To analyse structure**.
- **To analyse hydrodynamics**.
- **To design and improve production processes**.

Simulation also has its reason for being in the world of plant operations.

As will be seen in the section dedicated to *Digital Twin*, simulation can be an important ally in a shipyard, contributing decisively in the optimization of resources, both at the level of human resources and infrastructure as well as the rapid adaptation to new projects.

It is important to emphasize the issues related to plant safety: simulation models will provide the necessary information to be able to decide which will be configuration of the facility that could provide the optimum results for both:

- **Production**: definition and optimization of workflows in physical spaces, as in *figure 3*.

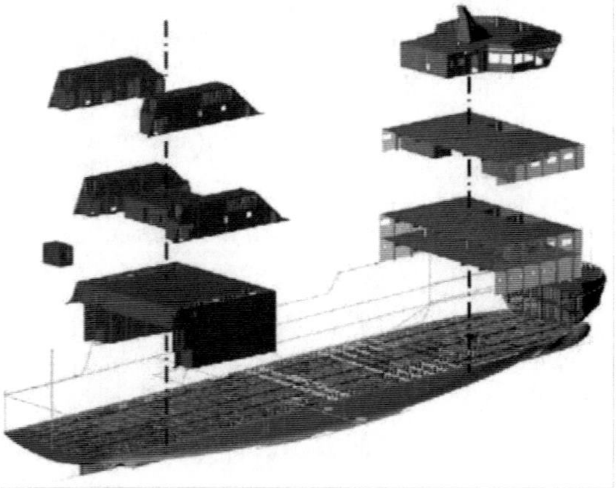

Figure 3: Simulation of structural blocks

- **Safety requirements**: distribution of spaces to ensure a proper evacuation in case of emergency and the definition of the *Evacuation Plan*.

A *Shipyard 4.0* that implements technologies, such as the use of autonomous machines, requires a particularly detailed analysis of the spaces, workers interactions, assets location and autonomous vehicles' paths. For this purpose, simulation tools become fundamental to analyse and provide solutions for the difficulties that could be involved due to the coexistence of autonomous vehicles and workers into the same working space.

A key factor is the constant progress of developments in the world of videogames and the potential that it is demonstrating to have in the world of the engineering.

These *game development platforms* natively offer access to a virtual world that includes physics, materials, animation, etc. allowing the creation of photorealistic environments in which to have the *Digital Twin* of the product and the shipyard, as well as the workers and vehicles that interact in it. Once done, there is free way for simulation in this virtual environment.

With these tools, it will be possible to analyse the different distribution and operation's options into the virtual world and test them under different conditions, like for example:

- **Workers and autonomous vehicles**.
- **Different environmental conditions**.
- **Failure of some systems**.
- **Product assembly operations**.
- **Facilitate evacuation process**.

The creation of the connection between the *3D CAD model* of the product and the infrastructure in which the product is manufactured, and the game development platform, opens the door to endless possibilities, in addition to the simulation itself. In many cases, the core of the products that offers the implementation of *Digital Twin* and the access to *Virtual Reality* and *Augmented Reality* technologies is based on a Video Game development platform. In the paper *The use of CAD Systems to manage modularity in multi-role warships* (*Pérez and Péter, 2019*) it is detailed how *CAD* tools could help in the design of this kind of warships when we combine simulation and a concept that can be called *variants*.

If a *CAD* or *PLM* system includes the possibility to work with *variants*, the complex case of the design and construction of multi-role warships can be faced in better conditions. In that case, the importance of the simulation of each variant before taking any decision is a key factor.

5. ARTIFICIAL INTELLIGENCE

Control system in the ship can include some *Artificial Intelligence* predictive processes integrated in the bridge overall control system, which helps to deduce the consequences of maintenance operations, from doing in the correct time as well as delaying or skipping them (*Pérez et al., 2018*).

This procedure applied is including in the edge/fog computing methods, due to only involves the ship inner communication network, delaying the massive download of operation data upon arrival at port. During navigation data transfer should be only applied to critical operations.

Navigation data can be also useful in design and production phases, to correct some processes in order to obtain more efficient systems, and more efficient designs. This is only possible applying some *Artificial Intelligence* process to this data, classifying, processing and getting some results.

This working methodology, to be deeply profitable for both actors, requires a joint venture between ship-owner and shipbuilder.

Artificial Intelligence processes based on navigation data, in the multi-boat paradigm, can obtain information to improve design and production processes, which can be applied to the current series, or an evolved variant of this vessel type, or other ones (*Benayas-Ayuso and Pérez, 2019*).

Ship operation phase is not the only one which produces a set of *Big Data* to be processed by an *Artificial Intelligence* System, in the production phase, some calculus can be done in the workshops or even delegated in a cloud system, to be distributed. This data, in *CAD Artificial Intelligence* tools, can be classified generating working sequences, design automatic checks, and automatic design processes.

At this point, some sceptic people can think: *Artificial Intelligence* is going to substitute designers work, but *Artificial Intelligence* is going to augment the capabilities of this designer, making work less stressing and more efficient.

Artificial Intelligence applied to a *CAD* System should be based on the standard ways to control a design; rules based design, lessons learned and cognitive rules.

First on the list, rules based design, it is the shipyard standards book of rules. These rules are the base of any shipyard design, and are the first one to be learned by our *Artificial Intelligence* System. These technology can be used for the definition of 3D model in a *CAD* system, as shown in *figure 4.*

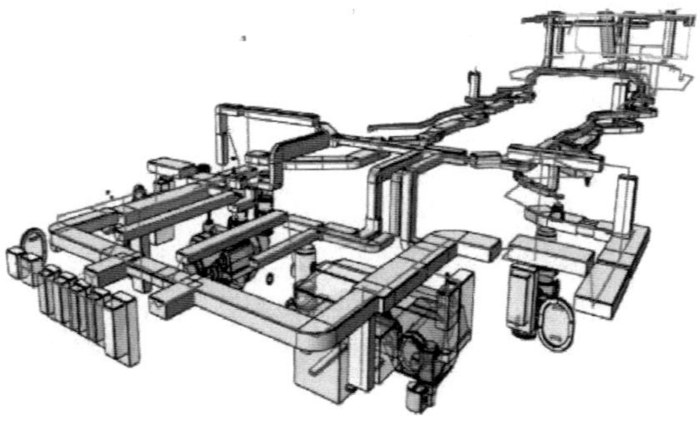

Figure 4: Electrical design 3D arrangement

To improve it, some other technologies can be applied, like Cognitive *Artificial Intelligence*, which is the base of natural language processing in an *Artificial Intelligence* solution. This cognitive *Artificial Intelligence* can be run over the shipyard standards book of rules, and with some user help, try to improve understanding in application of the rules from the *Artificial Intelligence* tool. All *Artificial Intelligence* solution requires a specific time to learn the correct ways to apply the set of rules.

Next step, also based on natural language spelling, can be add the *lessons learned* to our *Artificial Intelligence* solution, creating a mixed ecosystem of rules to be applied to the current design process.

And last step, but not less important, *Cognitive Rules*, these rules are deduced from the current design and also from production error input in the *CAD* System, like design incidences.

6. DIGITAL TWIN

Digital Twin is a concept that means the connection between the real asset and the design, the virtual world, as seen in the *figure 5.*

Figure 5: Idealistic representation of the Digital Twin

Digital Twin means a continuous and bidirectional exchange of *Data* between the real asset and the model, with the objective to have a real-time synchronization between those worlds. *Real Asset* to *Virtual Data* flow will be possible with a combination of sensors sending information through the communication infrastructure and human operators sending data by using *Apps* normally designed for devices that allow mobility and wireless connection. Human or automatic processes, if the virtual asset includes *Artificial Intelligence* or some automatic program that simply has an output depending on some input parameters, can manage both communication flows. At the basis of the *Digital Twin*, the model is mandatory.

There are different ways to have it:

- **Spherical photographs**: it is a *360º* horizontal and *180º* vertical photograph that can be reproduced in an interactive way into a *3D* representation.

- **3D scanning**: This method is widely explained in the paragraph that talks about *3D printing*.

- **3D *CAD* model**: This is the best option for a new infrastructure, or for a facility that already has the *3D model* that can be easily updated to current situation and be used as *Digital Twin* without excessive effort.

The combination of *3D CAD model* and *3D scanning* is perhaps the most interesting. With both technologies, it is possible to measure, and this is a very important advantage. Shipyard is not a static scenario, that is why the best option is a *Digital Twin* based in *3D models* generated with a *CAD* system and when necessary with models generated by using *3D scanning* technologies by transforming the clouds of points in volumes and surfaces.

It is important to assume that a fully integrated *Digital Twin* is not always reachable and it will be frequent to find partial implementations that consists just in data acquisition from machines in order to analyse and use for different purposes, but once reached this first step, it will be easier to move in the *Digital Twin* direction.

Augmented Reality application in Industry could be wide, and in part also offers many of the possibilities that *Virtual Reality* proposes but it is necessary to distinguish in each case what is the best option. *Augmented Reality* has a strong link with reality; this doesn't happen the same way with *Virtual Reality*, but if this *Virtual Reality* is used over *Digital Twin*, the link is stronger. This link with reality consists on the exploiting and visualization of the information generated by *IoT*. *Augmented Reality* allows as well to overlap the information generated by the *IoT* infrastructure directly, see history, trends, forecasts, warnings... Never-ending possibilities that *Digital Twin*, through *Virtual Reality* could also offer remotely, being able in a future, who knows, to reach a completely remote management of a complex infrastructure interacting with *Digital Twin*.

7. CONCLUSIONS

As an example of *IoS*, the connectivity in smart ships will be extended to the commercial mission to act autonomously in operation conditions. A commercial vessel can transmit its navigation situation or its loading conditions. All these means a huge amount of information to be managed and analysed. New programs have to be developed to obtain the best use of such information so that the design can be improved from real function information of the design and it can be self-maintained with the connection with this huge cloud information to create method that the objects can achieve certain intelligence.

The growth of the *IoS* is linked to the increase of information and the management of *Big Data*, with the property that somehow *IoS* identifies information, direction, and order to a specific purpose, while the concept of *Big Data* is more generic.

Last advantage is, if something happens in a ship, it can be reproduced in the shipyard based on this model, and getting a full condition simulation. This helps to focus the problem.

There is another important piece of the puzzle: the security, veracity and trust in the data transactions between all the different participants. *Cloud* and *security* are two pieces of the game and without them, all this ecosystem will never work. Data transactions should be guaranteed and all the steps should be tracked from the first step to the last one of the process thanks to *blockchain*.

In the middle of this ecosystem we will find a *3D model*, created mainly with *CAD* tools and this means that the importance of *CAD* will be even more capital than today. The interface between the *3D model* and the rest of the *Industry 4.0* is the *Digital Twin*, the link between the reality and the virtual world, both living concurrently during the evolution of the construction and extending this world of possibilities to the entire lifecycle of the product.

CAD and *PLM* tools, as an important part of all this environment, should also evolve to be easily linked to these technologies, but also it is important to be adapted to the new generation of users that demands a different kind of interfaces and workflow and this will require an important effort.

This big world is now open and Shipbuilding is starting in it, but the potential is clear and to remain competitive is mandatory to study how these technologies can improve the benefits by upgrading processes, resources, workflow, cooperation between stakeholders… the tools are here, now it is necessary to study and analyse our particular case. There is no global solution for all industries, not even a global solution for a particular industry, but digitalization is on hand and it is mature enough to start implementation and be part of our strategy for the future.

8. REFERENCES

[1] BENAYAS, A. and PÉREZ, R. (2019). *What should shipbuilding expect from the CAD/CAM systems of the future?*. The Naval Architect magazine. April. pp. 28-31.

[2] MUÑOZ, J.A. and PÉREZ, R. (2017). *CAD tools for designing smart ships in the world of the Internet of Things*. SMART SHIP 2017, 24 and 25 January.

[3] PÉREZ, R.; BENAYAS, A. and PÉREZ-ARRIBAS, F. (2018). *Data management for smart ship or how to reduce machine learning cost in IoS applications*. SMART SHIP 2018.

[4] PÉREZ, R. and PÉTER COSMA, E. (2019). *The use of CAD systems to manage modularity in multi-role warships*. Warship 2019.

INSTALLATION INNOVATION FOR FLOATING OFFSHORE WIND

A P Crowle, P R Thies, University of Exeter, UK

ABSTRACT

Floating offshore wind turbines are an emerging technology with notable prototype and demonstration projects being installed in recent years. Future floating wind farms enable the access to significant new wind resources in expanses of water too deep for conventional, bottom-fixed farms. Spar, barge, TLP and semisubmersible types have been deployed as demonstration units. Pre commercial units have been installed off the coasts of Portugal and the East Coast of Scotland.

This paper discusses naval architecture requirements for the temporary conditions of construction, ocean transport and installation. The requirements of Marine Warranty Surveyor guidelines for temporary conditions and the Classification Society requirements during the temporary phase are also taken into account. It is expected that major offshore maintenance will be carried out on floating offshore wind turbines by towing them back to more sheltered port locations. This paper is a literature and guideline review with additional information on intact stability during the tow to the offshore location. Thus, marine operations play a pivotal role throughout all phases of a floating wind farm's life cycle. In particular, uncertainties associated with offshore installation work can extend construction schedules and increase the capital expenditure required for the project. This paper will be of interest to researchers and practitioners in the floating wind sector, considering innovative installation methods as a means to reduce overall project lifecycle cost.

NOMENCLATURE

[Symbol]	[Definition] [(Unit)]
υ	Kinematic viscosity (N s m^{-2})
ρ	Density of water (kg m^{-3})
P	Pressure (N m^{-2})
AHT	Anchor handling tug
COG	Centre of Gravity
CUM	Cubic metres
Deg	Degree
DP	Dynamic positioning
FOWT	Floating offshore wind turbine
GM	Metacentre above COG
HAZID	Hazard indentification
HTV	Heavy Transport Vessel
HVDC	High voltage direct current
LOA	Length overall
M	Metre
SPMT	Self propelled modular transporter
SSCV	Semi submersible crane vessel
T	Tonnes
UXO	Unexploded ordinance
WFO	World forum offshore wind
WTIV	Wind turbine installation vessel

1. INTRODUCTION

There are several floating offshore wind turbine (FOWT) types in various stages of design. Only two types are currently producing pre-commercial electrical power. Figure 1 depicts the four main types namely barge, semi-submersible and Tension Leg Platform(TLP). This paper considers the installation aspects from port to offshore location. These temporary phases of a floating offshore wind turbine face many engineering challenges. As turbine sizes and distances from shore increase, weather becomes more severe and more important due to longer transit periods, increasing the logistics challenges for prospective floating wind farm developers and operators.

Figure 1 Floating options (courtesy ciercoenergy)

Marine operations play a pivotal role throughout all phases of a wind farm's life cycle. Uncertainties associated with offshore installation can extend construction schedules and increase the capital expenditure required for a given project. Installation costs can account for approximately 30% of the overall project cost and it is anticipated that informed engineering decisions in this area present further cost saving potential, through innovation. The increasing remoteness and heightened weather conditions increases the complexity of the marine operations.

There are 6 floating offshore wind turbines in UK waters (with another 4 under construction). Bottom fixed offshore wind turbines currently depend on wind turbines installation vessels (WTIV), which are limited to about 60m water depth though some WTIV are being modified to 80m water depth. Floating wind turbines are not so constrained bt water depth because they are completed in port or in a sheltered harbour.

There are about 60 different concept designs for floating offshore wind, two of which have reached pre commercial status, namely the Hywind Spar and Principle Power semi submersible. A further eight have been deployed at sea as demonstrators.

Floating wind turbines have their turbine nacelles and blades installed in a port or sheltered harbour. It is the intention to carry out maintenance on barge and semi submersible types back in a port.

The structure of the paper is as follows, section 2 reviews the current status of floating wind options, section 3 describes the temporary phases during installation, section 4 outlines the applicable rules and guidelines whilst section 5 discusses potential innovations. Section 6 provides a brief discussion and conclusions are drawn in section 7.

2. CURRENT STATUS OF FLOATING WIND

2.1 Types of Floating wind

This section will briefly review the main floating wind concepts and describe their main characteristics regarding stability and towout.

2.1.1 Spar

Spars have been derived from spars used in Oil and Gas. For in place stability, they require solid ballast to be placed in the base, a marine operation carried out afloat, figure 2. They require a large water depth 70 to 100 for construction and tow out.

Figure 2 Hywind Towout (courtesy Equinor)

2.1.2 Semi submersible

The semi submersible types are also derived from oil and gas technology, figure 3. For in place intact stability they depend on a large second moment of inertia water plane area, buoyancy and water ballast. They are fitted with large heave damping plates, under the columns, which means they take up a large area on land or dry dock or alongside an outfitting quay.

Figure 3 Semi submersible (courtesy Principle Power)

2.1.3 Ring barge

Ring barges, figure 4, are sized to dampen motions by sizing their moonpool. In place stability is derived from large buoyancy.

Figure 4 Concrete Ring Barge (courtesy Ideol)

2.1.4 TLP

The TLP is derived from oil and gas platforms, figure 5. In place stability is from taut vertical mooring lines or tethers. TLPs have low intact stability during tow to site.

Figure 5 TLP (courtesy Bluewater)

2.1.5 Suspended weight

The suspended weight option, figure 6, derives its in place stability once the weight has been deployed offshore. Tow out intact stability is low.

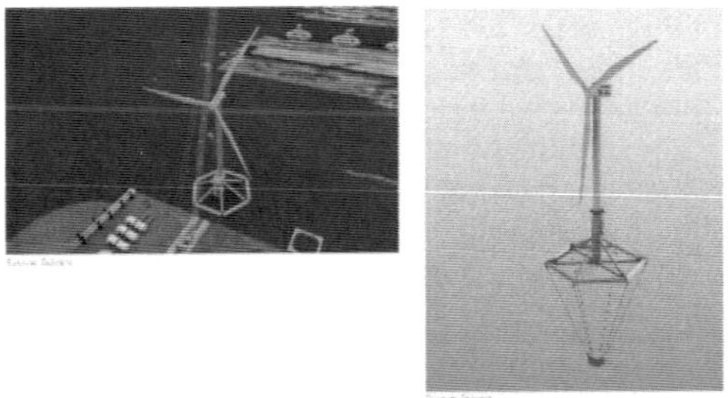

Figure 6 Suspended weight (courtesy Saipem)

2.2 Status

Floating wind has matured, enabling small-scale prototype projects and the world's first pre-commercial arrays. Floating wind farms are the key to opening up new wind resources in expanses of water too deep for conventional, bottom-fixed wind farms.

This makes it an important area of research for the industry, Ref [1]. But as with many new technologies, one of the biggest obstacles standing in the way of full-scale floating wind commercialisation is cost. The turbines are of similar cost for both fixed and floating offshore wind structures. Installing anchors for floating offshore wind turbines is similar in time to installing the substructure for a bottom fixed wind turbine. However industrialized manufacturing of the floating platforms has not taken place so they are significantly more expensive than bottom-fixed offshore wind power. Floating wind still lags far behind bottom fixed wind in terms of commercial readiness due in part to lack of port facilities for construction and turbine outfitting. Compared to monopiles, substructures for floating wind turbines are, for now, considerably more expensive to manufacture and assemble. Steel substructures are many times heavier and more labour-intensive to put together than a simple jacket or tripod or monopile for a bottom fixed structure. However it is expected that in large water depths beyond say 80m that floating offshore wind turbines become a more realistic option.

Attaching the turbines to their substructures is one of the areas where floating wind has a clear cost advantage over bottom-fixed: turbines can be installed in a much more controlled environment, and without the use of expensive jack-up vessels. However, spar technologies require deep, sheltered waters and offshore cranes, resulting in a turbine installation process that is more costly than semi-submersibles and TLPs.

While more vessels are required for floating wind installation compared to bottom fixed substructures, these are considerably cheaper to charter than a jack-up vessel. An exception would be in the case of TLPs that, if not self-stable in towing, require bespoke installation vessels, which would incur significant expenditure.

Trial projects like the WindFloat project in Portugal and Scotland and Hywind in Scotland have yielded progress ref [2]. However, floating wind still has not reached the same status as bottom-fixed wind in terms of commercial readiness. Innovation is needed in port facilities and offshore installation to reduce the length of installation schedules and costs.

3. TEMPORARY PHASES

3.1 Phases

For a floating offshore wind turbine the temporary phases start with loadout from land onto a heavy transport vessel, or float-out from a dry dock, and conclude when the moorings and subsea cables have been connected. In addition the temporary phases include seabed surveys at the offshore site, along the tow route, the cable routes and at the ports used for construction.

The following temporary phases are required:

- Seabed Survey,
- Anchor and mooring foundation installation,
- Substructure construction
- Turbine construction
- Completed structure tow to offshore site
- Pre-lay Grapnel Run before cable laying
- Subsea Cable Installation
- Cable Burial

Seabed surveys include bathymetry, soil conditions, wreck appraisal and unexploded ordnance (UXO) identification. Some elements of the temporary phase include return to port for major maintenance or demolition.

3.2 Weight

It is vital that weight control is used throughout the project life to ensure that the floating offshore wind turbine floats at the required draft during the different temporary phases. Weight control should be performed by means of a well defined procedural system, such as that described in Ref [6]. Weight control procedures shall be in operation throughout construction and outfitting when afloat.

A detailed weight control procedure is required at the start of design. The weight report should consider and include:

- Net dry weight
- Gross dry weight

- Installation weights for loadout, dry tow, in port, wet tow, in place
- Gross installed weight
- Centres of gravity for all conditions
- Radii of gyration for temporary and permanent conditions

The substructure will be weighed before loadout to measure weight and the plan centre of gravity.

A trim and stability book is to be produced covering:

- Hydrostatics
- Ballast tank capacities and centres
- Allowable VCG curve
- Damage conditions

An inclining experiment may be carried out either on the substructure or completed structure at the outfit port. This will be used to determine the vertical centre of gravity.

3.3 Substructure construction

3.3.1 Spar

To date 6 steel Spars have been installed in Europe, though the demonstration project has now been decommissioned. The installation procedure, starting at the shipyard is as follows.

- construct substructure horizontally in a shipyard
- loadout with self propelled modular transporter (SPMT)
- submersible heavy transport vessel (HTV)
- ocean transport to a deep water site
- lay out temporary inshore mooring
- float off
- upend
- connect to inshore mooring
- add permanent solid ballast
- fit turbine
- tow offshore
- connect to prelaid mooring
- add permanent water ballast
- connect to pre laid power subsea cables

Figure 6 Loadout of Spar by SPMT (courtesy Navantia)

Innovation was required for the lifting frame design used by Saipem to lift and install the turbine, figure 7. The lifting frame had been developed from a moored floating sheer leg crane lift of a complete turbine onto a fixed structure.

The innovation required on Hywind by Saipem included se of their semi submersible crane vessel (SSCV):

- use of dynamic positioning
- lift onto a floating structure
- efficient damping system
- robust guiding method
- strong mating flange

, Figure 7 Lift of turbine onto Spar (courtesy Saipem)

3.3.2 Semi built on land

To date three, pre commercial semi submersible FOWTs have been completed and another five are under construction. The construction procedure on land and subsequent installation can be summarised in the following sequence:

- construct substructure vertically on land
- loadout with self propelled modular transporter (SPMT)
- submersible heavy transport vessel (HTV)
- ocean transport, figure 8
- float off at the turbine assembly yard
- moor to quay
- fit turbine components
- pre commission
- tow offshore
- connect prelaid mooring
- permanent water ballast
- connect to pre laid cables

Figure 8 Substructure dry transport (courtesy Boskalis)

3.3.3 Semi built in a dry dock

To date three, semi submersible FOWT has been constructed in a drydock and installed in the following sequence:

- construct vertically in a dry dock, figure 9
- fit temporary buoyancy to minimise draft
- tow out of dry dock
- tow to the turbine assembly yard
- moor to quay
- remove temporary buoyancy
- fit turbine components
- pre commission
- tow offshore
- connect to pre laid mooring
- permanent water ballast
- connect to pre laid power cables

Figure 9 Semisub in drydock (courtesy Principle Power)

There are limitations on using dry docks because of the width of the structure, including the heave plates.

3.3.4 Barge in a dry dock

To date one, concrete barge FOWT has been constructed in a drydock and installed in the following sequence:

- build on a pontoon barge
- move into dry dock
- separate substructure from pontoon barge
- tow out of dry dock
- tow to the turbine assembly yard
- moor to quay
- fit turbine components
- pre commission
- tow offshore
- connect pre laid mooring
- permanent water ballast
- connect to pre laid cables

3.3.5 TLP built on land

No TLP FOWT are currently operating. The small Blue H demo has been decommissioned. Construction is as follows:

- construct vertically on land
- loadout by SPMT onto a submersible HTV
- ocean transport to the turbine assembly yard
- float off
- moor to quay
- fit turbine components
- fit temporary buoyancy
- pre commission
- tow offshore
- connect and tension vertical mooring
- remove water ballast
- remove temporary buoyancy
- connect to pre laid cables

Figure 10 0.08MW TLP (courtesy Blue H)

3.4 Tow to site

All the FOWTS types are towed from the outfit yard to the offshore locations using an anchor handling tug (AHT). A further AHT accompanies the tow.

A dedicated installation vessel with work class remotely operated vehicle (ROVs) is used to connect the pre laid mooring lines to the FOWT. Tensioning of the mooring lines is done by a winch on one of the AHT.

A specialised cable installation vessel is required to connect the subsea power cables

3.5 Anchor installation options

The main types of anchor under consideration are listed in table 1. Drag anchors and small suction piles can be installed with a anchor handling tug. Large suction piles and driven piles require medium sized crane vessels for installation plus anchor handling tugs for pre laying the mooring lines on the seabed.

Table 1 Anchoring options

	Barge	Semi-Submersible	Spar	TLP
Drag anchors	Suitable	Suitable	Suitable	Not suitable
Suction piles	Suitable	Suitable	Suitable	Suitable
Drive piles	Not preferred because of noise from pile driving and its effect on ocean mammals			Suitable

Chain mooring lines are connected to the anchor and pre laid on the seabed, prior to the arrival of the floating wind turbine.

3.6 Duration

The estimates of installation durations below, table 2, assume the same offshore location, assuming tow distance of about 290 nautical miles from the outfit port to the offshore site speed of 3 knots. These estimates are based on similar oil and gas structures taking account of different shapes and sizes.

Table 2 Typical installation durations

Item	**Barge Days**	**Semi Days**	**Spar Days**	**TLP Days**
Anchor installation	12	6	6	18
Substructure loadout onto a Heavy Transport Vessel	2	2	2	2
Dry tow to outfit port	5	5	5	5
Float off from Heavy Transport Vessel to a sheltered harbour.	2	2	4	2
Outfit turbine tqower, nacelle and blades in a sheltered harbour	10	10	10	10
Tow to site	4	4	4	4
Connect moorings	12	6	6	18
Connect power cables	5	5	5	5
Total	**52**	**40**	**42**	**64**

4. RULES AND GUIDELINES

4.1 Classification

The following classification societies have developed design requirements for floating offshore wind turbines:

- ABS ref [9].
- Bureau Veritas ref [10].
- DNV ref [11].
- Class NK ref [12].

They provide guidance on all temporary phases, including load-out, transportation, installation, maintenance and repair.

The focus of the classification societies lies on the design and construction of the permanent works, as well as their integrity for all temporary phases.

The installation procedures documented in the installation manual are to be provided to the society for information. Installation procedures are to be submitted to an appropriate third party for review, to ensure compatibility with the site conditions and the FOWT design. The installation manual may include:

- personnel qualifications and skills
- interface points and any required technical specifications for civil and electrical construction works
- specialized tooling and required lifting fixtures or equipment
- limiting environmental conditions
- quality control check points, measurements and inspections, required by the design
- installation loads and load conditions
- description of safety instructions and planned environmental protection measures
- quality recording and record keeping processes

Installation tolerances are to be specified in the installation procedures, and duly taken into account in the design calculations.

The installation survey includes, but is not limited to, the following operations:

- installation of anchors
- deployment of mooring lines
- test loading of anchor and lines
- connection to floating platform and tensioning
- post-installation inspection of the system

4.2 Marine warranty surveyor

The marine warranty provides certificates to proceed for:

- loadout
- substructure transport
- float off
- tow to offshore site
- mooring and subsea cable connection

The approval certificates input is:

- weather forecast for the intended marine operations
- use of best industry practices
- suitable personnel
- design for temporary works
- construction of temporary structures
- marine vessel status
- installation equipment suitability
- risk analyses
- contingency procedures

Design calculations include:

- weight control
- ballasting
- intact stability
- damage stability
- motions
- global strength
- local strength
- rigging sizes

Further details are given in ref [7] and ref [8].

5. INSTALLATION INNOVATIONS

Installation starts with the loadout of the floating offshore wind turbine substructure onto a heavy transport vessel or the floatout from a drydock and finishes with the mooring line and subsea cable connection. The industrialisation of floating wind requires specialised construction yards and inshore floating cranes or large land based cranes.

Floating technology ref [5] is used to deploy offshore wind in sea areas that have high wind speeds but are too deep to be cost effective for bottom fixed foundation installation. There is innovation potential in this area to support the commercialisation and future deployment in new deep water, high wind speed resources. There is a significant expected growth in the floating offshore wind market, in the UK, wider Europe, and internationally. Innovations in this sector will include dynamic high voltage cable systems, moorings for challenging seabed conditions as well as for very deep and shallow water, foundations and new installation, maintenance and fabrication innovations.

As floating turbines get larger and are being situated in deeper, rougher waters, reliable and cost-effective moorings are becoming essential for deployment increases to occur.

The development of new, cheaper to produce and install substructures, as well as validation of existing designs, are essential to lower the cost of floating wind. The layout of construction ports and outfit ports need to avoid constraints on layout space and carne capacity.

The development of commercial scale projects of approximately 500MW or greater is required to demonstrate the feasibility of floating wind as a competitive energy source.

Innovations in wind turbine floating substructure design as well those that decrease the costs of manufacture and installations will be the main drivers of deep-water wind farm deployment. Significant innovation opportunities exist for substructure, installation, logistics as well as offshore operations.

The fabrication of a large quantity of floating structures must be given particular attention. Floating structures could be built in modules with each module being built in specialized areas and by specialized suppliers. That would allow a 'serial-like' fabrication process to be put in place achieving optimum quality.

In terms of final assembly due to the nature of the floating solutions, there does not seem to be a suitable alternative to assembly at port. Thus, large investments need to be made in order to increase the port area, or drydock sizes, and dredging at the fit out quay.

Urgent innovation is also needed for offshore maintenance, ref [3]:

- Heavy lift maintenance operations from floating vessels offshore.
- 3D motion compensation systems from floating crane vessels when working close to floating wind turbine.

The tow to port maintenance innovation work should seek to:

- Investigate the procedures for disconnecting and reconnecting floating wind units in a large scale wind farm
- Evaluate key challenges and identify solutions to mitigate risks and costs
- Carry out intact and damage stability analysis
- For TLPs investigate how temporary buoyancy can be re-connected

6. DISCUSSION AND CONCLUSIONS

The stability triangle is useful to classify and locate the different floating technologies regarding their main stability mechanism, namely ballast, buoyancy and mooring. The triangle drawn in ref [13] considers in place conditions. Following the explanations above, we modify the stability triangle for tow out condition, see figure 10. Stability is measured by metacentric height (GM) and range of intact stability.

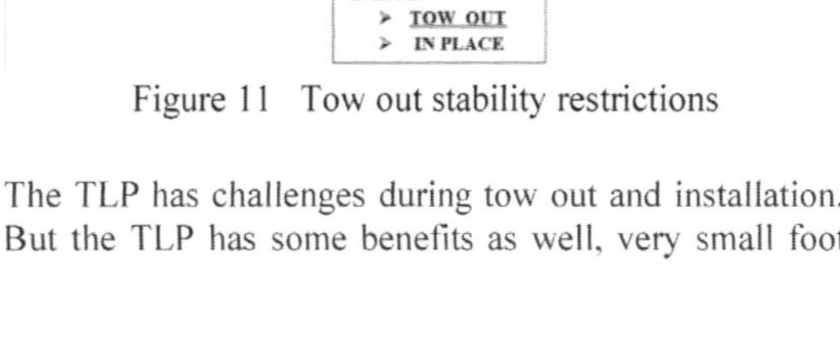

Figure 11 Tow out stability restrictions

The TLP has challenges during tow out and installation. But the TLP has some benefits as well, very small foot print, low structure weight, and less heave / surge motions of the turbine. Research on TLP installation is needed and may result in a requirement for new installation vessels.

Spars need deep water inshore for upending of the substructure and fit out of the turbines. Inshore moorings need to be laid for the Spars. Assuming parallel working then multiple inshore moorings are required. Thus, Spars need be constructed in locations where deep water is available close to the shore, such as Norwegian fjords.

Regarding semi submersibles from an inshore construction point of view their constructability varies with the location of wind turbine, figure 12. Some semi submersibles have the turbine over a corner so that the operating radius of the onshore crane is as small as possible. Other semi submersible options have the turbine in the centre, but have no underwater bracing between columns and in this case, the onshore crane can work from the corner of a fit out quay.

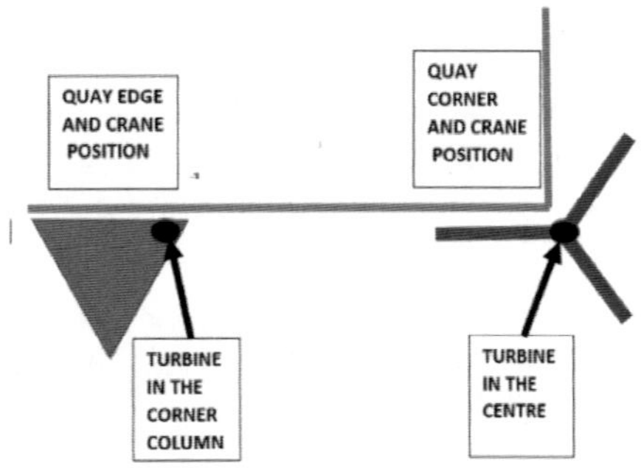

Figure 12 Semi submersibles at fit out quay

7. CONCLUSION

In general, a greater number of conventional vessels are required for floating wind installation compared to fixed offshore wind turbines. An exception is the installation of TLPs that, if not self-stable in towing, require bespoke installation vessels or large temporary buoyancy tanks, which would incur significant expenditure.

There is still considerable development needed to reach a stage of mass production for floating offshore wind turbines. Innovation in installation will assist in bringing forward large floating offshore wind farms.

8. ACKNOWLEDGEMENTS

Alan Crowle thanks his colleagues at the University of Exeter for their assistance in preparing this paper. PR Thies would like to acknowledge the support throughout the EPSRC Supergen ORE Hub [EP/S000747/1]

9. REFERENCES

1. Pre-Catapult – Re-Energise Winter 2017
2. Ore-Catapult – Re-Energise Winter 2020
3. Carbon Trust Phase III floating wind summary 2020
4. Wind Europe Floating Offshore Wind Energy A Policy Blueprint For Europe
5. Carbon Trust Energy Innovation needs assessment – Offshore wind, 2019
6. ISO 19901-5 Weight control
7. ISO 19901-6 Marine operations
8. DNVGL-ST-N001 Marine operations and marine warranty
9. ABS Guide For Building And Classing Floating Offshore Wind Turbines July 2020
10. Bureau Veritas Classification and Certification of Floating Offshore Wind Turbines January 2019
11. DNV Rules For Classification Offshore Units DNVGL-RU-OU-0512 Edition October 2020 Floating Offshore Wind Turbine Installations
12. Class NK Guidelines for offshore floating wind turbine structures, July 2012
13. Stability of Floating platforms (Thiagarajan, K. &Dagher, H. (2014))

10. AUTHORS BIOGRAPHY

Alan Crowle is a naval architect working for a research degree at the University of Exeter, into the installation floating offshore wind turbines. He has a BSc (1974) from University of Newcastle-upon-Tyne and MSc (2020) from the University of Plymouth.. He has over 50 years marine experience in design, construction and offshore installation. His work includes steel jackets, semisubmersibles, FPSOs, large offshore and onshore modules, near-shore pipelines and LNG facilities. He has design experience in the installation of bottom fixed ofshore wind turbines and fixed High Voltage Direct Current offshore platforms.

Prof Philipp R. Thies is an Associate Professor in Renewable Energy in the College of Engineering, Mathematics and Physical Sciences (CEMPS) at the University of Exeter. His research interest lies in the reliability engineering of renewable energy technologies with a focus on offshore energy. In current projects, he is Principal Investigator on a Joint Industry Project (Carbon Trust) for Floating Offshore Wind to demonstrate Intelligent Mooring Systems. He leads Exeter's contribution to the €45million EU Interreg initiative 'Tidal Stream Industry Energiser Project (TIGER)' and is PI on the Innovate UK funded project "Autonomous Robotic Intervention System For Extreme Maritime Environments (ARISE2)", led by L3 Harris. Prof Thies is a Co-Director of the EPSRC Supergen ORE Hub [EP/S000747/1] and Co-Investigator in the EPSRC/NERC Centre for Doctoral Training in Offshore Renewable Energy (IDCORE) [EP/S023933/1], training the next generation of offshore engineers.

AN ANALYSIS OF THE EFFECTS OF ALTERNATIVE FUELS ON A 68M SUPERYACHT

H M Maidment, V Boccolini, Y Wang, L A Wright, Solent University, UK

SUMMARY

This paper seeks to analyse the impact of alternative fuels in the design of a modern Superyacht. The fuels, selected as zero net greenhouse gas emissions across their lifecycle, would allow the yacht to comply to the newest IMO targets but would certainly imply some changes and modification to the design. These modifications are analysed under different aspects and in relation to different parameter, in order to have a clear and objective analysis; the effects of the Class rules are also evaluated where possible and applicable. The conclusions presented leave some questions still open as there are no unique solutions that the designer could take but the pros and cons of the different solutions are presented and the constraints that they impose on the new designs clearly identified.

1. INTRODUCTION

The International Maritime Organisation (IMO) has set an ambitious target for the reduction of greenhouse gas emissions produced by shipping industry. The main aims of the IMO strategy are to reduce [1]:

- CO2 emissions by 40% by 2030.
- CO2 emissions by 70% by 2050.
- Greenhouse gas emissions by 50% by 2050.

A solution to achieve these targets is a switch to using fuels which have drastically lower emissions or which have near or actual net-zero emissions. As a subsection of the global fleet, the superyacht sector is moving towards cleaner solutions, with the availability of new fuel technologies enabling this segment of the market to adopt newer solutions.

The authors provide an engineering impact analysis on three possible new fuels, Biodiesel, Renewable Methanol and Hydrogen on a case study of a 68m motor superyacht. This takes into account the possibilities and complexities of refuelling, storage, energy transference methods and regulatory requirements with specific focus on the issues surrounding volumetric density, mass density and stoichiometric ratios.

2. OBJECTIVES

The authors assess the impact of biodiesel, renewable methanol and hydrogen on the design of a superyacht.

The study focuses on those areas which will see direct modification of the design and main characteristics of the vessel. In particular, keeping the autonomy constant, the propulsion and powering options were analysed; considering actual at market and likely near future technologies. As direct consequence of the different type of fuels, the effect on the tankage and the limitation that they could bring were evaluated under different scenarios. The scenario analyses were defined considering the effect of the applicable regulation and guidance that some Classification Societies have already issued and, where possible, already established practice for the use of the particular fuel.

3. THE CASE STUDY YACHT

The case study yacht considered in this study is a 68m explorer superyacht designed for a range of 3500nm. The yacht uses a hybrid diesel-electric propulsion system, which uses conventional diesel engines and a traditional powertrain to reach cruising and maximum speed and an electric motor to reach, at max speed of 7kn, remote or environmental protected area. The yacht main characteristics and data and the General arrangement are reported in Appendix **Error! Reference source not found.**.

4. POSSIBLE ALTERNATIVE FUELS

This report provides the in-depth analysis of Biodiesel, renewable methanol and hydrogen as possible alternative fuels, in-lieu of traditional marine diesel oil. Each fuel was assessed considering the effects it would have on the design of the case study yacht, including on:

- Refuelling
- Storage
- Propulsion and energy generation
- Regulatory requirements

A typical marine diesel oil system was used as a benchmark with the three alternatives evaluated against this system.

Biodiesel is considered as possible alternative because it does not bring any technical problems to replace diesel. It is produced from biomass, for example agricultural wastes, waste cooking oils, or soybeans.

Hydrogen is a possible transitional fuel that can be used both in internal combustion engines (ICE) and in fuel cell in case of hybrid or fully electric solutions. A standard ICE requires modification to utilise hydrogen as well as the fuel systems. Wärtsilä is developing ICE engines that could run on hydrogen [2]. The GHG emission of hydrogen is zero if the fuel is produced by renewable energies (so called 'green hydrogen'); where hydrogen is formed from fossil fuels (e.g. from methane steam reforming (MSR)), the majority of the current market, life cycle emissions can be higher than traditional diesel systems. Similarly, methanol can be used in ICE and fuel cell systems. Methanol can be produced using methane from fossil sources, to be considered renewable and zero-emission, so-called Biomethanol, is produced by gasification of organic materials followed by conventional methanol synthesis.

4. FUEL COMPARISON

4.1 REFUELLING

4.1 (a) Conventional marine diesel

Diesel fuel is pumped into a yacht at ambient temperature and pressure using a conventional pump.

4.1 (b) Hydrogen

All hydrogen storage methods present technical issues to refuelling. To increase available energy and subsequently storage capacity, hydrogen needs to be compressed. Due to the high diffusion rate, hydrogen will readily leak to the atmosphere if a full seal is not maintained. Hydrogen is highly combustible and potentially dangerous in the case of a fuel leak.
As a state-of-the-art technology, hydrogen has few applications and the regulatory framework for the refuelling is quite often demanded to local authorities [3]. However, the following solutions may be considered:

- Filling compressed hydrogen tanks at a production facility and then swapping the tanks with those on board. This would simplify the refuelling process in all locations.
- Filling compressed hydrogen tanks at a production facility and delivering the fuel aside ship ready to be pumped at high pressure and/or compressed onboard.
- Cryogenically cooled liquid hydrogen delivered using a similar method to traditional diesel, with the additional consideration of cryogenically cooling all pipework.

4.1 (c) Biodiesel

Biodiesel presents no technical issue in relation to refuelling and the same process used for conventional diesel is applicable [4].

4.1 (d) Renewable methanol

Methanol is a liquid fuel under atmospheric pressure and temperatures below 64˚C [5]. Development in the road transport industry proves that methanol fuel tanks can be refilled using the same method as conventional diesel. Methanol compatible tanks and pipes are required [6].

4.1 (e) Refuelling assessment

Biodiesel and Methanol use conventional refuelling methods although methanol does require the fuel system to be adapted to resist its corrosive characteristics.
Hydrogen cannot be refilled using conventional methods unless all piping is cryogenically cooled and insulated. The most feasible refuelling method is to replace all tanks and have them refilled at a production site. It could be possible to deliver compressed fuel alongside the yacht, decompress it, fuel at atmospheric pressure and then recompress at higher pressure on board but this process would require a considerable amount of energy and a dedicated compressor area on board.

5. STORAGE

5.1 CONVENTIONAL MARINE DIESEL

The case study yacht stores diesel fuel in tanks amidships and aft. A range of 3500nm is achieved by storing 109,000l for the engines and generators. The total fuel mass is 90.85 tonnes.

5.2 BIODIESEL

Biodiesel has similar characteristics to conventional diesel; no modifications are required to convert the storage system. However, Biodiesel has a calorific value between 39 and 41 MJ/kg compared to 43 MJ/kg of diesel which would therefore require increasing the volume to 111,360l and the mass to 97.62 tonnes.

5.3 HYDROGEN

There are three primary methods of storage:

- Compressed gas storage
- Cryogenic storage
- Cryogenic compression storage

Hydrogen has the highest energy density of common fuels by weight but the lowest by volume at 143.0 MJ/kg and 0.0108 MJ/L respectively under atmospheric conditions [7].

5.3 (a) Compressed Gas Storage

Hydrogen gas can be compressed between 20 and 80 MPa using composite storage cylinders. Details of two typical compression pressures are summarised in Table A

Pressure (MPa)	35	70
Gravimetric Density (wt%)	5.5	5.2
Volumetric Density (MJ/L)	2.5	5.0
Storage Efficiency (%)	56.5	54.2

Table A - Compressed Hydrogen Storage Summary [7]

Increasing the cylinder pressure increases the volumetric density by a factor of two, while only decreasing the gravimetric density by 0.3% due to the increased tank wall thickness [7].

5.3 (b) Cryogenic Storage

Cryogenic storage of hydrogen requires cooling to below hydrogen boiling point of 20°K (-253.15°C) to a liquid form. This process increases the volumetric density of the fuel, requiring a reduced volume over compressed hydrogen methods. The process is energy intensive requiring 2.8 MJ/kg to liquefy the hydrogen, three times greater than is needed to compress the fuel to 70 MPa.
This storage method suffers from boil-off losses, requiring insulation to avoid this. Current tanks use double wall insulation with several metallic foil layers separated by glass wool; the space between inner and outer vessels is evacuated, creating a vacuum. Tanks without this insulation absorb heat from its environment causing the liquid hydrogen to boil which increases the pressure inside the vessel, leading to possible rupture or explosion. [7]

5.3 (c) Cryogenic Compression Storage

The cryogenic compression storage method includes pressurising liquid hydrogen, further increasing the liquid density from 70g/L to 87 g/L at 24MPa and volumetric density of 15 MJ/L assuming an increase of a multiple of 3 compared to compressed hydrogen [7]. This method enables high storage capacities and eliminates the issues of boil-off losses. Any losses vent to maintain constant pressure.

5.3 (d) Hydrogen Comparison

The comparison of different hydrogen storage systems is summarised in Table B and Table C. They include the weight of fuel required to maintain the range of 3500nm assuming constant energy storage equivalent to diesel. These calculations do not account for boil-off allowances since it varies with storage type, volume, number and shape of tanks.
At a stoichiometric air-to-fuel ratio of 34:1 produces power between 85% and 115% that of gasoline, however the combustion temperature is extremely high increasing the NOx emissions. To reduce the NOx emissions a ratio of 68:1 is typical, causing a power reduction of about 50%. To account for the power reduction, the fuel storage volume is doubled when compared to gasoline engines [8].
In this study only the potential of cryogenic compression is discussed as the currently low technological maturity causes installations to have a limited efficiency but enabling an insight into future capabilities.
As seen from Table B, hydrogen weighs up to 56% less than diesel fuel. However, hydrogen requires at least double the storage volume due to its low volumetric capacity to maintain the range.
Table C shows the weight of hydrogen, including the storage system, is at least 4 times greater than diesel. Hydrogen must have large tanks installed which can withstand high pressure and/or cryogenic cooling, decreasing the gravimetric capacity of the fuel. Table B and Table C show a trade-off between weights, volume and range when considering fitting these systems to yachts.

Fuel	Diesel	Compressed Hydrogen 35 MPa	Compressed Hydrogen 70 MPa	Cryogenic Hydrogen	Cryogenic Compression (Potential at 24 MPa)
Volumetric Capacity (MJ/L)	35.00	2.50	5.00	8.00	15.00
Volume of Fuel (m3)	109.2	1,528.80	764.40	477.75	254.80
Volume accounting for power loss	109.2	3,057.60	1,528.80	955.50	509.60
Density (g/L)	832.0	17.60	26.30	52.60	87.00
Fuel only Mass (t)	90.85	53.81	40.21	50.26	44.34

Table B – Summary of Hydrogen Fuel Masses [7]

Fuel	Diesel	Compressed Hydrogen 35 MPa	Compressed Hydrogen 70 MPa	Cryogenic Hydrogen	Cryogenic Compression (Potential at 24 MPa)
Gravimetric Capacity (wt%)	-	5.50%	5.20%	5.11%	9.20%
Total Mass with Storage (t)	95.40	978.43	773.22	983.55	481.90

Table C – Summary of Hydrogen System Masses [7]

5.4 RENEWABLE METHANOL

Methanol has 45% volumetric energy density of conventional diesel; 15.9 MJ/L compared to 35 MJ/L resulting in a required volumetric increase of 120% and mass increase of 109%.
Methanol is a corrosive liquid, so methanol-compatible coatings are required to enable the use of current fuel tanks [9].
Due to the low flashpoint of methanol, arrangements for purging of fuel tanks and piping using nitrogen is required [9] [10] [11].

5.5 STORAGE ASSESSMENT

Hydrogen increases the fuel storage volume significantly, requiring between 7.2% and 43.4% of the yacht's interior compared to 1.5% required for diesel to maintain the 3500nm range using ICEs.
Methanol is highly corrosive, so all tanks require protective coatings or use of stainless-steel materials and all pipes need replacing with double piped systems. The use of methanol doubles the amount of fuel required, needing 3.4% of the yacht's volume compared to 1.5% for diesel.
Biodiesel requires the minimal amount of change from conventional diesel. No modifications to the system are required and a fuel volume equal to 1.6% of the vessel's volume is required, an increase of 0.1%.

6. ENERGY GENERATION

6.1 CONVENTIONAL MARINE DIESEL

Marine diesel is combusted within ICE transferring chemical energy into shaft work and heat.

6.2 HYDROGEN

6.2 (a) Hydrogen Comparison

Hydrogen fuel can be used with current ICEs, fuel properties are shown in Table D.

Fuel Property	Diesel	Natural Gas	Hydrogen	Gasoline
Density (kg/m^3)	850-1000	0.65 gas 422.8 liquid	0.08 gas 70.1 liquid	730
Calorific Value (MJ/kg)	40-43	50	120	45
Flammability Limits (%)	0.6-5.5	5.3-15	4-75	1.4-7.6
Auto-ignition Temperature (˚C)	340-350	540	585	277
Diffusion Coefficient (cm^2/sec)	0.05	0.16	0.61	-
Minimum Ignition Energy (MJ)	0.24	0.29	0.02	0.24
Flame Velocity (m/s)	-	0.38	1.85	0.37-0.43
Adiabatic Flame Temperature (˚C)	-	1941	2207	2307

Table D – Comparison of three types of fuels [12] [13]

Modifications to the air-to-fuel ratio and intake temperature in ICEs can prevent preignition and engine knocking [12].
Hydrogen has a low ignition energy and high flame speed; at a stoichiometric ratio of 34:1 these properties can cause pre-ignition. This can be prevented through increasing the ratio to 68-parts air to 1-part fuel; lowering the combustion temperature and decreasing the power produced by 50% [8]. This also has the effect of increasing the thermal efficiency. Methods such as pressure boosting, turbocharging and direct injection systems have been shown to increase the power produced [13].

6.2 (b) Hydrogen compatible fuel cells

Fuel cells are electrochemical devices that convert chemical energy into electrical energy with efficiencies of up to 80% [12]. Different fuel cell technologies can be used depending on the purity of the fuel. Phosphoric Membrane Fuel Cells (PMFCs) have lower efficiencies and operate at higher temperatures (≈180˚C) than Proton Exchange Membrane Fuel Cells (PEMFCs) but are capable of withstanding poisoning by carbon monoxide from less pure hydrogen. PEMFCs are used with pure hydrogen fuel at temperatures between 80 to 100˚C. The effect of reducing the power density by 50% is not experienced within fuel cells since there is no combustion and no NOx emissions.

6.3 BIODIESEL

Biodiesels can be used in ICEs with no modifications.
The reduced HHV of biodiesel (39-41 MJ/kg) compared to conventional diesel (43 MJ/kg) [4] suggests power decreases by up to 10% for the same volume of fuel, an increased fuel consumption is expected to maintain power. A 1.98% volume increase is suggested in Appendix 0.

6.4 BIODIESEL

Methanol is compatible with current ICEs. It exhibits a greater peak efficiency of 43% and maintains efficiencies over 40% over a wider range of speeds and loads compared to diesel [5]. MAN [10] have created a dual fuel engine for use with methanol, using double walled pipes to capture eventual fuel leaks. These pipes are naturally ventilated and monitored for methanol leakage.
Gong et al. [14] and Li et al. [15] researched the effect of injection and ignition timings with methanol in modified diesel engines. Both found that diesel engine's injection and ignition timings could be modified and optimised to suit the new fuel being used. This is different for each engine series and, in case, for any fuel blend.

6.5 RENEWABLE METHANOL COMPATIBLE FUEL CELLS

Direct Methanol Fuel Cells (DMFCs) have been developed specifically for use with pure methanol and is a PEMFC, see Spiegel [16].

6.6 ENERGY GENERATION ASSESMENT

All fuels may be used with ICEs to convert the chemical energy to mechanical energy. Conventional diesel, biodiesel and methanol may be used with ICEs with little or no change to the fuel consumption although methanol requires modifications to the system to prevent leakage and corrosion. Hydrogen may be used in ICEs although hydrogen's stoichiometric ratio causes a 50% reduction in power from the same amount of fuel.
Using fuel cells with hydrogen allows the energy to be transferred more efficiently, equalling and surpassing the efficiency of all the other fuels in ICEs. Direct Methanol Fuel Cells have been developed and have a low operating temperature.

7. REGULATORY REQUIREMENT

This section analyses the major aspects of rules for classification in regards of each fuel. Rules for commercial ships have been analysed since alternative fuels have not been applied to yachts or accounted for in classification at the time of writing. All requirements are in addition to those required in SOLAS regulations.

7.1 CONVENTIONAL MARINE DIESEL

Rules for conventional marine diesel were used as a baseline [17].

7.2 HYDROGEN

The primary changes for using hydrogen and their impact on the design of the yacht are: [18]

- 1.5 x 2.5m air locks at entrances to hazardous zones
- 900mm A-60 rated cofferdams surrounding all fuel storage zones
- Compressed storage may not be stored below decks

7.3 BIODIESEL

No changes are required for biodiesel compared to conventional diesel.

7.4 RENEWABLE METHANOL

The main factor that will alter the design for the use of renewable methanol are: [19].

- Fuel tanks and pipes shall be at least 800mm from the ships side shell
- 600mm A-60 rated cofferdams surrounding all fuel storage zones
- Nitrogen purging systems
- All piping must be double walled and gastight

7.5 CONSIDERATION FOR FUEL CELLS

This section summarises primary changes from diesel impacting the yacht's design. Full details are provided in DNVGL [20].

The main considerations are:

- Fuel cell space boundaries shall be gas tight and designed to avoid accumulation of hydrogen rich gas
- Fuel cell spaces must be accessed from the open deck or through and air lock (see **Error! Reference source not found.**)
- 900mm A-60 rated cofferdams surrounding all fuel cell spaces

7.6 REGULATORY REQUIREMENT ASSESMENT

No modifications are required for biodiesel from conventional diesel.

Methanol does pose challenges to the internal layout and systems design. All tanks require 600mm cofferdams to ensure no risk of crew or passenger exposure. This will impose greatly on the space available on the lower decks and could cause functionality losses due to loss of volume.

All fuel system appliances would be replaced by treated and double walled variants to comply with regulations, imposing additional costs. It would also require a Nitrogen purging system, inducing engineering complexity and costs.

Methanol may be preferred since it occupies much less of the yacht volume than hydrogen, so would provide a GWP reduction while maintaining the requirements typically desired in a vessel of this type. Also, since hydrogen is typically compressed, all tanks will be fitted so they are open to the deck, causing the centre of gravity to rise making the yacht unstable.

8. IMPACT ANALYSIS

8.1 MASS AND VOLUME

As shown in Figure A, diesel is the optimal fuel with the lowest fuel mass and volume. The figure also shows the impact on the total mass and volume of the yacht in full load diesel displacement compared to the increase that the alternative fuel would require.

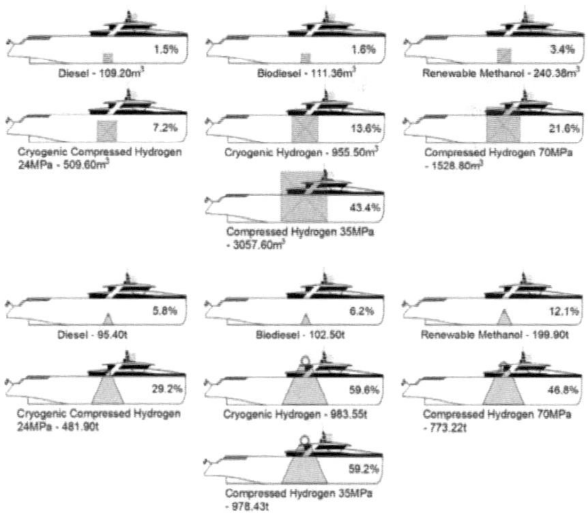

Figure A – Fuel mass and volume comparison with ICEs

8.2 STABILITY

Figure B shows the comparison of the stability of the yacht for the different fuels in full loaded condition.

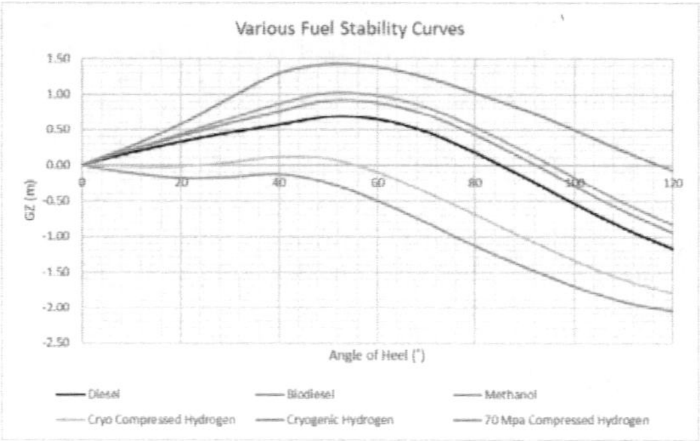

Figure B - GZ stability curves for all fuels

8.3 BIODIESEL

Biodiesel has a similar calorific value to conventional diesel and would cause a fuel mass increase of 7% and volume increase of 2%. Figure B shows the increased mass improves the yacht stability increasing the area under the GZ curve and increasing the angle of vanishing stability to 92°; it complies with REG-A stability regulations.

This fuel can be refuelled and stored using the same methods as conventional diesel and has similar regulatory requirements.

8.4 RENEWABLE METHANOL

Methanol has a calorific value nearly half that of conventional diesel, causing a mass increase of 110% and volumetric increase of 120%. As with biodiesel, the rise in mass increases the area under the GZ curve and angle of vanishing stability to 98°; it is compliant to all REG-A rules.

Methanol can be refuelled and stored using conventional methods however, both pipes and tanks require treatment to protect them from the fuel's corrosive characteristics and the possibility of leakages. To maintain the range and comply with regulatory requirements the tankage space would need to be expanded and separated from non-fuel spaces by 600mm cofferdams. This would greatly reduce the available space within the yacht.

8.5 HYDROGEN

Four storage methods of hydrogen are available and are categorised by compression and cryogenic storage. Compressive storage at 35 and 70 MPa increases the fuel mass by 978% and 773% respectively and volumetric increase of 2700% and 1300% compared to diesel. This can be refuelled through replacing the tanks when depleted. Compressed storage tanks cannot be stored inside the yacht's hull, so must be stored on deck or in a well-ventilated deck level compartment separate from living and recreational spaces, hence the replacement of the tank is a feasible refuelling possibility. This storage solution would rise the centre of gravity and consequently causes a loss of stability. In addition, it would also require 900mm cofferdams.

Cryogenic storage methods increase the energy density of the fuel, thereby increasing the mass by between 405% and 931% compared to diesel; the volume would be increased by 367% and 775%. No method of refuelling was found in literature. If non-compressed these systems can be stored below deck however 900mm cofferdams are required between any non-fuel spaces.

Cryo-compressed, cryogenic and 70 MPa compressed hydrogen were tested for stability since 35MPa compressed hydrogen would only increase the effect of 70MPa. As previously explained, any compressed storage method must be stored in a location open to the air. Therefore, the centres of gravity were taken as level with the upper deck. As shown in Figure B the yacht is generally unstable, therefore these storage methods are not advisable. The model was also tested for 70 MPa compressed hydrogen with a centre of gravity 1.5m (one deck) lower, returning positive but non-compliant stability curve as shown in Figure C. This simulation shows that there is a possible solution to comply with the rules and have a gas storage, but the solution needs to be engineered and the yacht designed around this area.

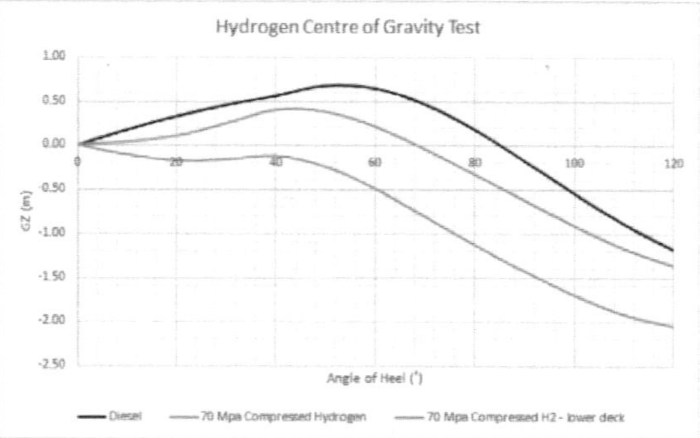

Figure C - GZ stability curves for hydrogen stored 1.5m below deck

Cryogenic hydrogen was the most stable storage option since it was stored below decks, achieving a very stable yacht with a moment of vanishing stability of 119.5°. However, it fails stability rules since the angle down flooding is vastly reduced compared diesel due to the increased fuel system mass. Also in this case, hydrogen could be used but the problems that it brings need to be addresses since the early stage of the design.

9. CONCLUSIONS

From an engineering perspective biodiesel and methanol are both promising alternative fuels for future use. The current technologies are well developed and can be adopted in yachts easily. Methanol requires the most modification and highest cost of the two fuels.

The analysis has proved hydrogen is not yet a suitable fuel for the yachting industry. The sample yacht would lose between 7.2% and 43.4% of the interior volume and would require a fuel system mass of at least 481.9 tonnes. The sacrifice required from the owner would likely be too much for the current technologies to be commercially viable. If these technologies were adopted, the yacht would need to be designed around the fuel system to ensure compliance with the current rules and regulations for stability and safety. It also needs to be noted that it is very likely that the systems on board would require more space due to the necessity of refuelling stations or pressure control.

In case of use of fuel cells, there could be the necessity of additional storage for batteries that has not been taken into consideration.

The technology is currently inadequate to permit hydrogen to be adopted in ICE so biodiesel and methanol are looking to be the best alternatives as transitional fuels.

This engineering analysis should be twinned with environmental impact analysis of the same fuels to gain a full insight into the future of yachts to better evaluate the impact of the life cycle of alternative fuels for Superyachts.

10. REFERENCES

1. INTERNATIONAL MARITIME ORGANISATION, "Reducing Greenhouse Gas Emissions from Ships," [Online]. Available: http://www.imo.org/en/MediaCentre/HotTopics/Pages/Reducing-greenhouse-gas-emissions-from-ships.aspx. [Accessed 8 September 2020].
2. WÄRTSILÄ CORPORATION, "Wärtsilä gas engines to burn 100% Hydrogen," Wärtsilä Corporation, 5 May 2020. [Online]. Available: https://www.wartsila.com/media/news/05-05-2020-wartsila-gas-engines-to-burn-100-hydrogen-2700995. [Accessed 27 November 2020].
3. OFFICE FOR PRODUCT SAFETY AND STANDARDS, "Regulations: Alternative Fuels Infrastructure," UK Government, 27 October 2020. [Online]. Available: https://www.gov.uk/guidance/regulations-alternative-fuels-infrastructure. [Accessed 27 November 2020].
4. N. YUSUF, S. K. KAMARUDIN AND Z. YAAKUB, "Overview on the Current trends in Biodiesel Production," Elsevier, Selangor, 2011
5. X. ZHEN AND Y. WANG, "An Overview of Methanol as an Internal Combustion Engine Fuel," Elsevier Ltd., Tianjin, 2014
6. R. L. BECHTOLD, M. B. GOODMAN AND T. A. TIMBARIO, "Use of Methanol as a Transportation Fuel," Alliance Technical Services Inc., Dunkirk, 2007
7. D. J. DURBIN AND C. MALARDIER-JUGROOT, "Review of Hydrogen Storage Techniques for on Board Vehicle Applications," Kingston, 2013
8. US DEPARTMENT OF ENERGY, "Module 3: Hydrogen Use In Internal Combustion Engines," December 2001. [Online]. Available: https://www1.eere.energy.gov/hydrogenandfuelcells/tech_validation/pdfs/fcm03r0.pdf. [Accessed 30 November 2020].
9. M. SVANBERG, J. ELLIS, J. LUNDGREN AND I. LANDALV, "Renewable methanol as a fuel for the shipping industry," Elsevier Ltd, Goteborg, 2018
10. MAN, "Using methanol Fuel in the MAN B&W ME-LGI Series," MAN, Copenhagen, 2014.
11. INTERNATIONAL MARITIME ORGANIZATION, "Methanol as Marine Fuel: Envionmental Benefits, Technology Readiness, and Economic Feasibility," International Maritime Organization, London, 2016
12. N. R. AMMAR AND N. F. S. H. ALSHAMMARI, "Overview of the Green Hydrogen Applications in Marine Power Plants Onboard Ships," International Journal of Multidisciplinary and Current Research, Egypt, Saudi Arabia, 2017
13. C. M. WHITE, R. R. STEEPER AND A. E. LUTZ, "The hydrogen-fueled internal cobustion engine: a technical review," International Journal of Hydrogen Energy, Livermore, 2005.
14. C. M. GONG, K. HUANG, Y. L. CHEN, J. L. JIA, Y. SU AND X. J. LIU, "Cycle-by-Cycle combustion variation in a DISI engine fueled with methanol," Changchun, 2011.
15. L. JUN, C.-M. GONG, S. YAN, H.-L. DOU AND X.-J. LIU, "Effect of injection and ignition timings on methanol," Elsevier Ltd, Changchun, 2010.
16. C. SPIEGEL, "Direct Methanol Fuel Cell Improvements," Fuel Cell Store, 27 March 2018. [Online]. Available: https://www.fuelcellstore.com/blog-section/direct-methanol-fuel-cell-improvements. [Accessed 6 August 2020].
17. DNVGL, Reciprocating internal Combustion Engines, January 2018 ed., DNVGL, 2018, pp. 8-85.
18. DNVGL, Propulsion, Power Generation and Auxiliary Systems, 2020, pp. 120-153.
19. DNVGL, Low Flashpoint Liquid Fuelled Engines, 2020.
20. DNVGL, Fuel Cell Installations, 2020.
21. Science Direct, "Cryogenic Hydrogen," 2020. [Online]. Available: https://www.sciencedirect.com/topics/engineering/cryogenic-hydrogen#:~:text=Cryogenic%20hydrogen%20has%20a%20density%20nearly%20twice%20that,which%20have%20provisions%20for%20cooling%2C%20heating%2C%20and%20venting.. [Accessed June 2020].

11. APPENDICES

11.1 SAMPLE YACHT MAIN DATA

Sample Yacht Characteristics			
Length Overall	68.00 m	Electric motors	2x 240 kW
Waterline Length	66.70 m	Batteries	164 units
Beam	12.50 m	Batteries Weight	10.25 t
Draught	3.56 m	Electric demand (ZEM)	1762.63 kWh
Displacement	1650 t	Fuel Tanks	109,200 l
Range	3500 nm	Emergency Fuel Tank	264 l
Maximum Speed	18 kn	Fresh Water Tanks	50,100 l
Cruising Speed	12 kn	Technical Water Tanks	86,500 l
Engines	2 x 1678 kW	Grey Water Tanks	12,600 l
Diesel Generators	2 x 565 kW	Black Water Tanks	3,660 l

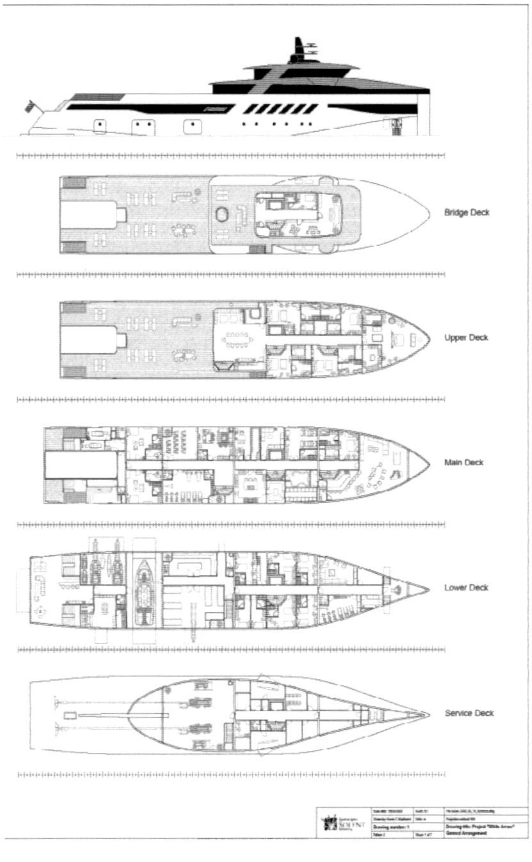

11.2 BIODIESEL VOLUMETRIC COMPARISON

Biodiesel volumetric comparison to conventional diesel maintaining a tank energy of 3,822,000 MJ

Fuel Type	MJ/l	Density (kg/m3)	Tank Volume (m3)	Annual Volume (m3)
Diesel	35.00	832.00	109,200.00	109.20
Biodiesel	34.32	876.60	111,363.64	111.36
	%age Volume Increase	Mass (kg)	%age Mass Increase	
Diesel	0.00%	90,854	0.00%	
Biodiesel	1.98%	97,621	7.45%	

12. AUTHORS BIOGRAPHY

Harris C Maidment is a recent graduate of the MSc Super Yacht Design course at Solent University. He completed the analysis of future fuels in relation to super yachts as his final project for this course. His previous experience includes working as a certification engineer at HPi CEProof.

Vittorio Boccolini holds the current position of Senior Yacht Engineering Lecturer at Solent University. He is responsible for lecturing UG program in Yacht Design and Masters programme in Superyacht design. He is also part of the research team working on the Implementation of Ship Hybridisation ISHY, an EU funded Project on ship hybridisation.

Dr. Yifan Wang holds the current position of Postdoctoral researcher at Solent University. He is part of the Implementation of Ship Hybridisation (ISHY) project and responsible for undertaking research in area of alternative marine fuels, contribute to advancing the field of life cycle assessment (LCA), life cycle cost assessment (LCCA) and multi-criteria decision making (MCDM). His previous experience includes PhD researcher in A Critical Examination of the Economic Value of Using LNG as an Alternative Marine Fuel.

Dr. Laurie Wright holds the current position of Senior Lecturer and Lead for Environmental and Engineering Research at Solent University. He lectures in Sustainability Management, and Life Cycle Assessment. He is leading a number of research projects, including the MarRI-UK funded Intelligent Ship Centre (IGNITE), and Solent's contribution to the Implementation of Ship Hybridisation (ISHY) and, the Ports Energy and Carbon Savings (PECS) an EU funded projects on ship hybridisation; and carbon reduction in ports.

COMPARISON OF EXPERIMENTAL TOWING TEST AND CFD ANALYSIS OF SUBMARINE MODEL ON THE SURFACE

T H Tun, Myanmar Maritime University, Myanmar

SUMMARY

Hull form, resistance and powering are need to consider in designing submarine. Submarine hull form is based on the five parameters. The submarine's surface resistance will be importance because of operating at high Froude Number. Model experiment is the most reliable data source on ship resistance determination due to the complication of the flow around the ship's hull. CFD simulation is also carried out for surface condition by FINEMarine and pressure distribution on the hull is appeared. Towing mechanism should be arranged into account in order to fix the pitch motion and to measure hydrodynamic forces. This paper briefs about the towing method and result between model test and CFD. It also includes comparison of wave formation for towing test and CFD.

NOMENCLATURE

D_H	Hull Diameter (m)
L_B	Bow length (m)
L_C	Cylinder length (m)
L_{CS}	Conical Stern length (m)
LOA	Length overall (m)
LWL	Length waterline (m)
R_t	Total resistance (N)
T	Draught (m)
V	Speed of vessel (m/s)
Δ	Displacement (tonnes)
γ	Conical Stern angle (degree)

1. INTRODUCTION

For designing submarine, Hull form, resistance and powering are need to consider. Model testing is carried out traditionally for estimation the resistance and power at the surface condition. The tested model is bare hull form.
Nowadays CFD tools are not only widely used for prediction of ship's resistance and power but also for other industries. Therefore, CFD solver is also used in design step and validation results is carried out by comparing the model test results. The model experiments were carried out in the Towing Tank at Marine Hydrodynamics Centre, Myanmar Maritime University [6]. Towing tank is 60m in length, 4m in breadth and 4m in depth and maximum carriage speed is 4 m/s.

2. OBJECTIVE

This paper analyses the resistance and compares the results with CFD simulation results for submarine bare hull. Towing mechanism is considered to test the model without running trim.

3. STUDY AREA

Operating the submarine at the surface condition increases the resistance. This case is one of the major problems in designing of submarine. Although the behaviour of submarine's resistance is as same as the normal surface ships, submarines have low freeboard that the majority of the hulls are below the water surface. At high Froude number, the wave making resistance becomes importance and generates higher waves.
To study this condition experimentally, it is necessary to design the towing scale model, resistance measurements can be recorded and by using appropriate scaling laws. Initially, model of 1.5 m in length is fabricated with wood and tested in model basin at Myanmar Maritime University. The CFD simulation is also carried out to compare the resistance and wave formation of the vessel.
The CFD solver can calculate with multi-phase flows and moving grids. In the multi-phase continuum, considering incompressible flow of viscous fluid under isothermal conditions, mass, momentum and volume fraction conservation equations can be written as [4] –

$$\frac{\partial}{\partial t}\int_V p dV + \int_S p(\vec{U} - \vec{U_d})\vec{n} dS = 0 \qquad (1)$$

where V is the domain of interest, or control volume, bounded by the closed surface S with a unit normal vector $\vec{n}$ directed outward. $\vec{U}$ and p represent, respectively, the velocity and pressure fields.
When the grid is moving, the so-called "space conservation law" must also be satisfied:

$$\frac{\partial}{\partial t}\int_V dV = \int_S \vec{U_d}\ \vec{n} dS = 0 \qquad (2)$$

4. STUDY APPROCH

4.1 DETERMINING THE MAIN DIMENSIONS

Due to the interaction between the propeller and the hull, a submarine bare hull form is based on the following five parameters [10] –

- The fineness ratio
- Prismatic coefficient
- Nose radius
- Tail angle and
- The position of the maximum section

Table 1 describes the dimensions of submarine model. Relation L/D is equal to 8.3 because of limitation of testing arrangement and model making by CNC. The design of vessel features an elliptical main hull and a conical stern as shown in Figure 1.

Table 1: Main Particulars of Model at Design Condition

Main Particulars	Unit
Overall length, LOA (m)	1.500
Hull Vertical Diameter, D_{VH} (m)	0.210
Hull Horizontal Diameter, D_{HH} (m)	0.180
Displacement, Δ (tonnes)	0.24
Bow length, L_B (m)	0.333
Cylinder length, L_C (m)	0.590
Conical Stern length, L_{CS} (m)	0.576
Draught, T (m)	0.172
Conical Stern angle, γ (degree)	20.2

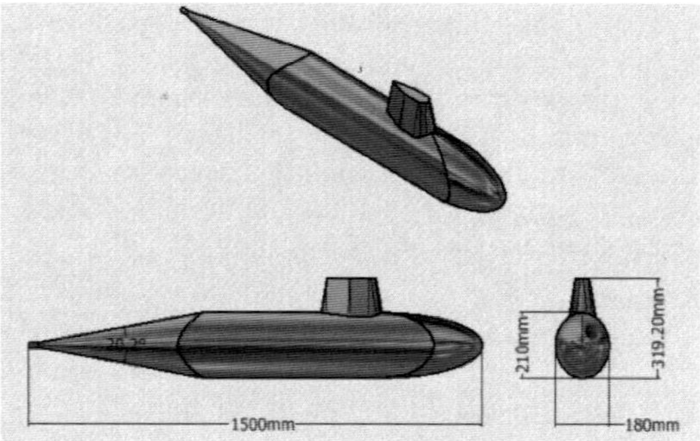

Figure 1 Model of submarine

4.2 MODEL MAKING AND TOWING MECHANISM

Fabricated wooden model without appendages is shown in Figures 2 below.

Figure 2 Fabricated wooden model

Submarine bare hull model is tested in towing tank at the surface condition and is observed that even keel condition cannot be controlled [12]. CFD simulation is also found out for that condition by FINEMarine and pressure distribution on the hull is shown in Figure 3. Therefore, the towing mechanism is designed to fix the pitching motion of the model as shown in Figure 4. The components of towing mechanism are listed in detail as the following Table 2–

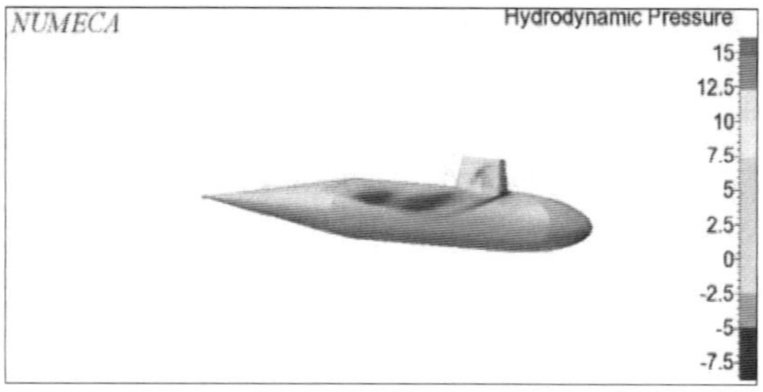

Figure 3 Hydrodynamic effect of the model

Table 2: List of Towing Mechanism

Symbol	Items	Function
	Load cell (sometimes called a strain gauge)	To measure the downforce or lift force
	Guide arm fit pins	To ensure the straight-line position on flow (2 numbers for bow and stern)
	Load cell (sometimes called a strain gauge)	To measure the resistance of model (attached to model)
	Vertical slipway or towing post	To measure the heaving motion and tow the model by carriage.

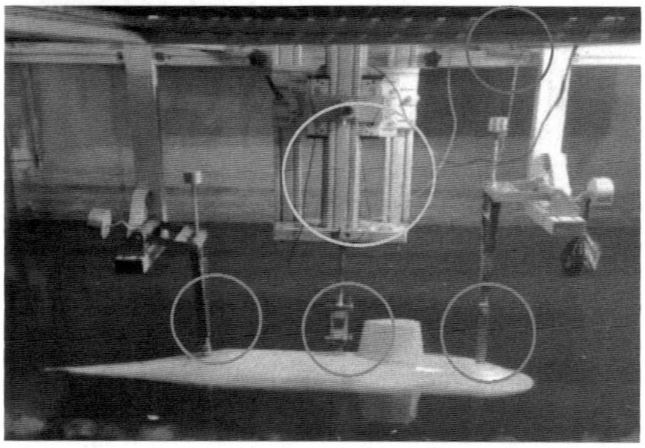

Figure 4 Towing mechanism arrangements

4.3 EXPERIMENTAL APPROACH

Experiments are tested for the speed range of 0.6 to 1.8 m/s and the resistance and down force are tabulated in Table 3 below -

Table 3: Model Resistance Test

Speed of Vessel V (m/s)	Total Resistance of Model R_t(N)	Down Force (N)
0.6	0.800	0.720
0.8	2.187	2.764
1.0	4.522	6.129
1.2	7.937	10.796
1.4	7.457	10.446
1.6	12.497	8.794
1.8	18.015	7.363

4.4 NUMERICAL APPROACH

The flow around the ship's hull is complicated and model experiments are still regarded as the most reliable data source on ship resistance determination; nevertheless, numerical methods have strongly advanced in this field, so that a combined use of both model tests and CFD codes can be very useful for ship design and for understanding the ship hydrodynamics [3]. CFD package is used to perform numerical computations: FINEMarine with the turbulence model k-omega (SST-menter). The initial mesh is X axis 24, Y axis 16, Z axis 12 and total cells are 843146 and the generated mesh of domain is presented in Figure 5.

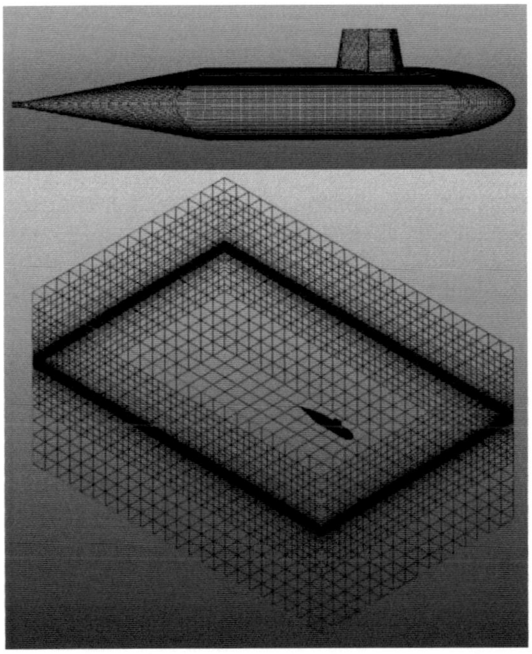

Figure 5 Generated mesh of domain

Towing test is carried out with fixed condition (locked for pitching motion). CFD simulation is also performed with fixed model for the fixed trim and sinkage. Figure 6 shows the wave formation of experiment and CFD simulation.

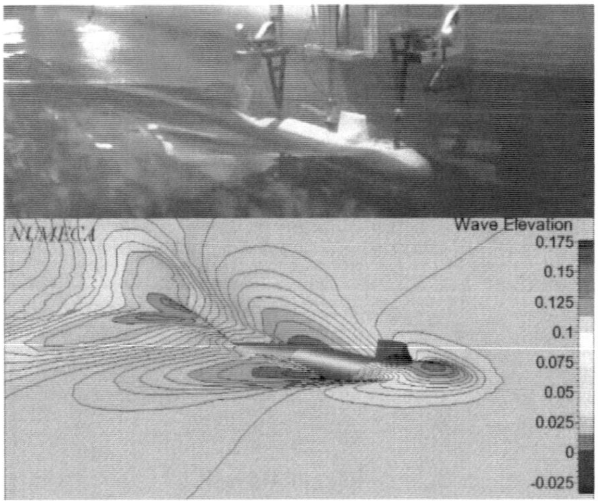

Figure 6 Comparison of wave formation Towing tank test and CFD with 1.8m/s

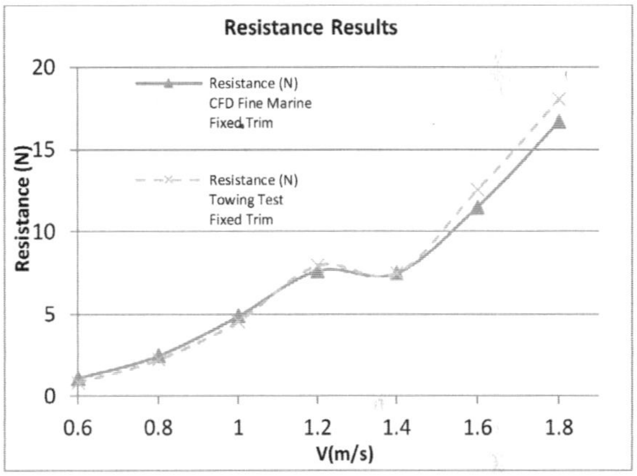

Figure 7 comparison of CFD and Towing test of ship resistance without appendages.

5. RESULTS COMPARISON

The submarine resistance is measured and wave forms are compared for the speed of 0.6 to 1.8 m/s. At the Froude's number of about 0.33, bow wave creates a trough at the stern and increases the stern wave system. This also increases the wave making resistance followed by resistance hump. The results are plotted in Figure 7. According to the results, it can be seen that the resistance of the vessel increases significantly from the vessel speed of 1.4 m/s. Although the resistance of the vessel becomes higher, the down force gradually decreases from the vessel speed of 1.4 m/s.

6. CONCLUSIONS

Towing test analysis for the model is experimented with fixed trim condition. Then CFD is also simulated with fixed model condition. The experimental test results are compared with CFD results and the comparison shows that they are well-matched for surface condition. To operate before Fn=0.33, hydrodynamic forces acting on the vehicle will become larger and will decrease value beyond Fn=0.33 (see in Table 3 Down force (N) column). So, it is still required to consider the hydrodynamic effect of the model in freely moving condition presented in Figure 3 and Table 3. After calculating the hydrodynamic effect of the model, hydroplane should be attached to ensure for even keel (no trim) condition.

7. ACKNOWLEDGEMENTS

The authors would like to express their thanks to Aung Khaing Min and Saw San Moe for all their support during the experimental model tests.

8. REFERENCES

1. DEY, A., 'A Study of a Two-dimensional Flow Around a Hydrofoil by Rankine Source Panel Method Using a Computer Program Based on C++', B.Sc. thesis, Dept. Naval Arch. and Marine Eng., BUET, Dhaka, Bangladesh, 2013.
2. CHOI, JE., MIN KS., KIM JH., et al., 'Resistance and Propulsion Characteristics of Various Commercial Ships Based on CFD Results', Ocean Engineering, 37: 549-566, 2010.
3. WATSON D.G.M., 'Practical Ship Design', Elsevier Science Ltd, 1998.
4. FINE™/Marine 9.1, Theory Guide.
5. FRANCESCO Capizzano, AMEDEO Migali, SALVATORE Miranda and CLAUDIO Pensa., 'Experimental and Numerical Study on the Efficiency of Trimaran Configuration for High-Speed Very Large Ship', 6th Symposium on High Speed Marine Vehicle (HSMV), Baia, Naples, Italy, September 2002.
6. ITTC 7.5-02-02-01, 'Recommended Procedures and Guidelines', Resistance Test, Revision 03, 2011.
7. KARIM, M.M. and NAZ, N., 'Computation of Hydrodynamic Characteristics of Ships Using CFD', International Journal of Materials, Mechanics and Manufacturing, Vol. 5, No. 4, November 2017.
8. NIZAM, M. S., 'Computation of Wave-making Resistance of Multihulls Using a Surface Panel Method', M.Sc. thesis, Dept. Naval Arch. And Marine Eng., BUET, Dhaka, Bangladesh, 2011.
9. KARIM, M.M. and Naza N., 'Computation of Flow Field Around Ship Hull Including Self Propulsion Characteristics at Varying Rudder Positions', 10th International Conference on Marine Technology, MARTEC 2016.
10. BURCHER, R. and RYDILL, L., 'Concepts in Submarine Design', Cambridge ocean technology series, 1995.
11. TARAFDER S. and SUZUKI K., 'Numerical Calculation of Free-surface Potential Flow Around a Ship Using the Modified Rankine Source Panel Method', Ocean Engineering; 35: 536-544, 2008.
12. TUN, T.H. and HTUN, Y.T., 'The Experimental Method and CFD Study of the Bare Hull Form of Underwater Vehicle at Near Surface Condition', ASEAN Engineering Journal, 11(1), Pp.24-33, 2021.

9. AUTHORS BIOGRAPHY

Thu Han Tun holds the current position of Professor / Head at Marine Mechanical Engineering Department, Myanmar Maritime University. He is responsible for academic management, research and development. His previous experience includes:

- the study of identifying design trends and the appendage drag influence on resistance performance
- Underwater Vehicles Design and Resistance Testing

INNOVATIONS IN FRICTION STIR WELDING FOR MARINE APPLICATIONS

S Cater, The Welding Institute (TWI), UK

SUMMARY

Friction stir welding (FSW) has been widely used for the fabrication of safety critical aluminium structures in the aerospace, road, rail and marine transport industries since its invention in 1991. Due to its ability to make high strength, tough, fatigue resistant welds with very low distortion, and the fact that the process can work underwater, considerable interest has been shown over many years in transferring the technology to the welding of steel. The difficulties in developing tools able to stir steel at the elevated temperatures required, in excess of 900oC, are now being overcome and the process is being developed for welding steel with an initial focus on marine applications. This paper discusses the technical advances being made at TWI and outlines the newly begun EU project "RESURGAM" which looks to develop the process for both the construction and repair of ships.

NOMENCLATURE

Exit hole	The hole left in the structure at the point where the tool is removed
FSW	Friction Stir Welding
HAZ	Heat Affected Zone
kN	kiloNewton
NDT	Non Destructive Testing
QA	Quality Assurance
SAW	Submerged Arc Welding
WPS	Welding Procedure Specification

1. INTRODUCTION

Friction stir welding is a solid-state welding process invented by TWI in 1991 and subsequently widely used for the fabrication of structures requiring high strength, lightweight, fatigue resistant joints. The process was originally developed for joining aluminium, as this is considered a difficult material to weld, and was subsequently developed for other hard to weld metals such as magnesium and copper. FSW was quickly adopted as a fabrication technique for spacecraft, trains, shipping and automotive components, as well as electronics assemblies and consumer goods, all applications where aluminium joints needed to be made that were strong, tough, fatigue resistant and lightweight. Once the good mechanical properties of FSW joints were recognised, along with the benefits of the potentially low cost and automated means of creating them, users began to request that the process be developed for steel too.

The development of tools sufficiently strong to stir steel, and to be chemically inert to the steel itself at the high temperatures required to soften it, meant that introducing FSW to steel was a challenging undertaking. However, small scale laboratory testing showed that the quality of friction stir welds in steel could be exceptionally good and thus the quest for industrially useful FSW tools for joing steel worthwhile. Such tools are now being introduced to the market and work underway to identify and exploit the many benefits of the process. An area where FSW in steel is likely to see early use is in shipbuilding and ship repair, two areas in which TWI is actively engaged to develop the technology as part of the RESURGAM project. This is a three year, EU funded, multi-national research initiative to develop the equipment, processes and qualification routes needed for fabricating ships and conducting underwater repairs of steel structures using FSW.

2. THE FSW PROCESS

2.1. HOW DOES FSW WORK?

Conventional fusion welding processes used for the fabrication and repair of marine structures are essentially a casting technique. The two components to be joined are brought together and energy applied, typically in the form of an electric arc or laser beam, to cause the metal at the region of the joint to melt and coalesce. As the energy source is removed this molten weld pool freezes and the previously separate components are now a single piece. When done well, this produces a strong, tough joint but there are potential problems, including:

- Structural distortion arising from the thermal energy input and the phase change from solid to liquid and back to solid;
- Changes in metal properties due to the creation of a cast microstructure in what was previously typically wrought plate;
- Changes in materials properties due to the addition of a filler material, or the boiling off or segregation of the original alloying elements;
- The potential for generating defects such as porosity, undercutting, lack of penetration and inclusions;

These issues can be overcome, but require the development and implementation of both specific technical measures and adherence to good working practices, e.g. Welding Procedure Specifications, along with appropriate quality control and post weld testing regimens.

In contrast, FSW is fundamentally a technique in which the components to be joined are forged together and never molten – akin to the way in which blacksmiths heat up steel and hammer it together! To make a simple butt joint using FSW, the two plates to be joined are placed side by side and the rotating tool is lowered and brought into contact with them at the joint line - as shown in Figure 1. The rotating tool generates frictional heating and this softens, but does not melt, the metal. As the metal becomes soft and plastic, the tool is driven into it until it is fully embedded to the tool shoulder, and then traversed along the joint line. The tool continues to heat and soften the metal immediately ahead of it, and the rotary action of the tool sweeps metal around it and across the joint line, uniting the two pieces of metal as one. At the end of the weld, the tool is extracted.

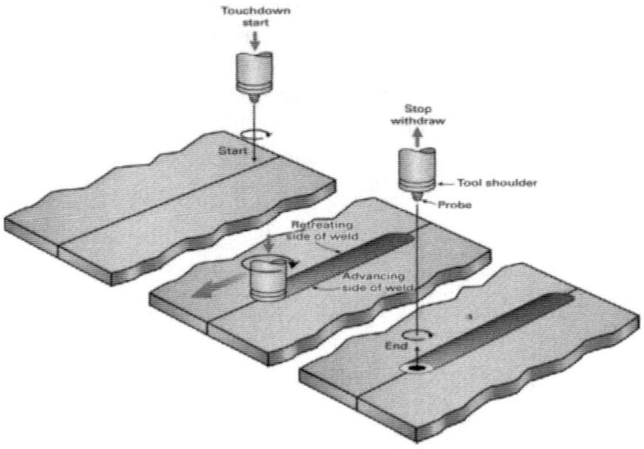

Figure 1. The general principle of friction stir welding.

2.2 BENEFITS & DISADAVANTGES OF FSW

The benefits of FSW arise primarily from:

- The elimination of melting in the joining process;
- The generation of a forged rather than cast weld microstructure;
- The elimination of human variability and skill from the welding process.

Considering these in more detail, the elimination of melting brings about a significant reduction in structural distortion, and enhanced health and safety benefits as there are no molten metal or fume hazards. Additionally, the process is far less energy intensive. The absence of a liquid metal weld pool also frees the process from gravitational effects and constraints: welds can be performed vertically (3G) and even overhead (4G) just as easily as a flat, horizontal weld. A further benefit is that FSW can also be performed under water, and has even been demonstrated to be effective when performed in oil. [1]

The mechanical, forging nature of FSW produces a fine grained, wrought microstructure rather than the more typical cast microstructure of fusion welding. This can be seen in Figure 2, which shows a multi-pass arc weld in 20mm thick carbon steel plate. The arc weld was made using a V preparation that was then filled by 38 sequential arc weld passes. These have produced a cast microstructure with large columnar grains and a large heat affected zone. A FSW tool was then passed through the upper portion of the weld, creating a very fine grained, forged microstructure with a much smaller HAZ. The microstructure in this FSW region is very similar to that of the wrought parent metal and, unsurprisingly, has very similar mechanical properties. [2]

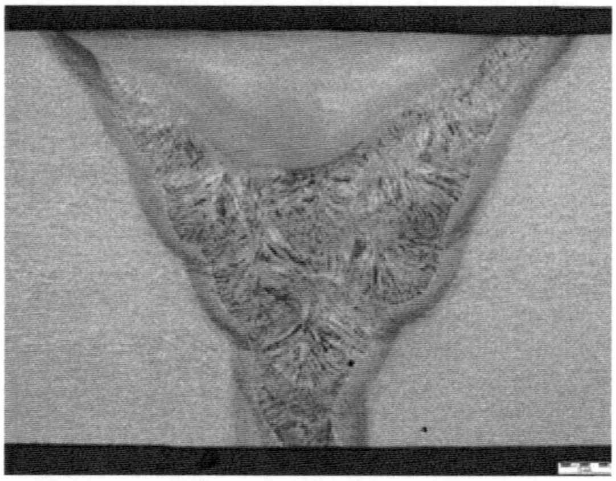

Figure 2. A metallographic section through a multi-pass arc weld in 20mm thick carbon steel, with the weld cap re-processed by FSW to show the difference in weld microstructure

Figure 3 shows a typical metallographic section through a friction stir weld in 6mm thick DH36 steel. The weld shows full penetration and no significant undercutting, with no visible internal defects such as voids, shrinkage cracking or segregation. The HAZ is small and the weld has a fine grained, equi-axed microstructure similar to the parent plate. As no filler metal is required for a FSW, the alloy composition remains unchanged and the costs of purchasing, storing and administering filler metals is eliminated from the welding process.

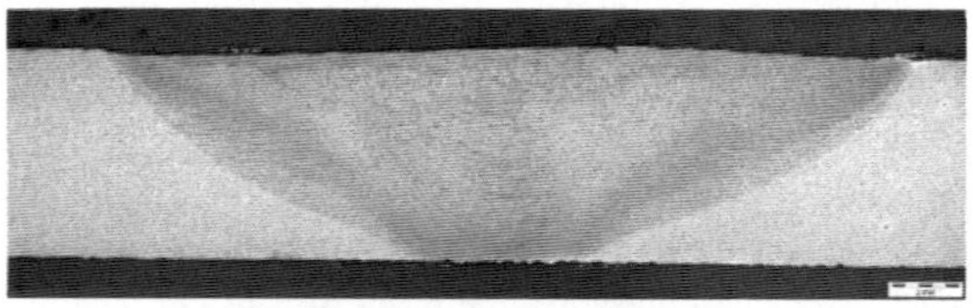

Figure 3. A metallograpghic section through a single pass FSW in 6mm thick DH36 steel.

Friction stir welding is a mechanised process, suitable for automation or robotisation. In its simplest form, it can be undertaken using a milling machine. The process is thus not dependent upon the skill of individual welders and so reduces process variability, allows for 24 hour operation, can be performed in hazardous environments and permits the use of in process statistical control and automated NDT / QA techniques.

Potential disadvantages of the FSW process are the requirements that the work pieces need to be firmly clamped to prevent them moving apart when the FSW tool is plunged into the joint line, and that the weld must be supported from beneath during welding to react the applied force of the friction stir welding tool. These are not difficult problems to overcome, but do need to be recognised. Thought also needs to be given to means of eliminating the exit hole left when the tool is extracted: this is typically achieved by using a run off tab.

2.3 WELD GEOMETRIES POSSIBLE

The physical, forging nature of FSW places some constraints on the weld geometries that can be achieved as the two faying surfaces are required to be brought into close contact and filler metal is not available to make up any significant gaps. That said, a wide range of joint geometries has been achieved in light metals with typical eaxamples shown in Figure 4.

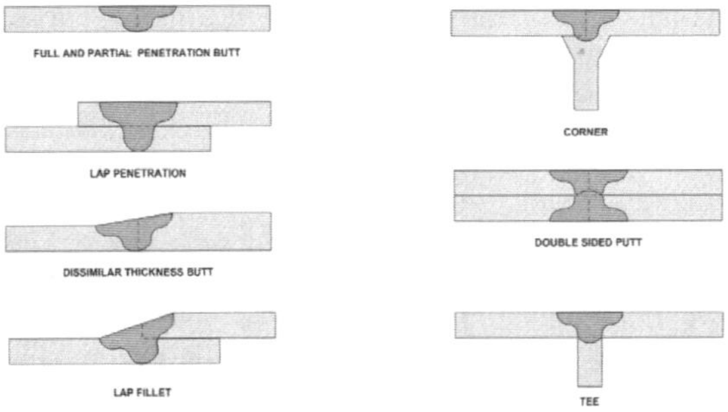

Figure 4. Joint geometries most easily achieved in FSW.

A variation of the FSW process, Stationary Shoulder FSW, has been developed and used in aluminium to allow corner and fillet welds to be produced, as well as permitting joints to be made between dissimilar thickness of metal.

An example of a radiused corner fillet weld made by this technique is shown in Figure 6. Here two 8mm thick plates of AA6082 have been joined using opposing passes of a FSW tool which simultaneously joined the plates and forged the radius into the plasticised metal before it cooled. Though the process is fully solid state, techniques are available (AdStir) to add filler metal to the weld if required to build up the fillet or add metal to compensate for poor fit up of the plates being joined.

In time, these techniques may be developed for welding in steel: any limitations are not imposed by the process itself but by the need to develop tools that are sufficiently strong and robust enough to stir the steel.

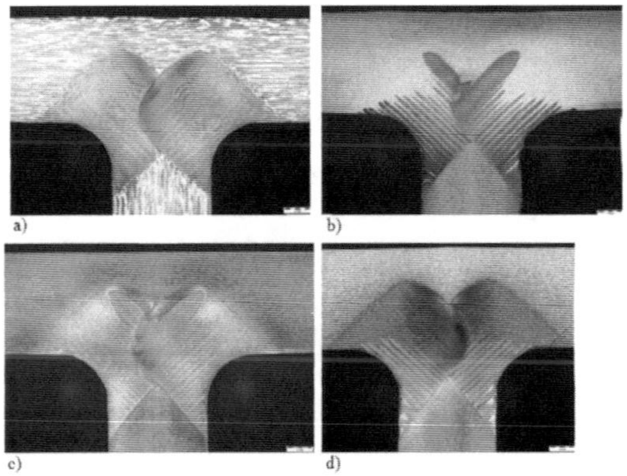

Figure 5. Metallographic section through a radiused corner fillet weld made by two FSW passes in 8mm thick AA6082.

2.4 MARINE USE OF FSW - ALUMINIUM

FSW has been a major enabler in the development of high speed, aluminium hulled vessels, particularly fast ferries and naval patrol craft. The welds have proven to be strong and stiff, with excellent fatigue and corrosion resistance. Figure 7 shows a number of naval craft which utilise FSW in their construction.

Figure 6. The USN LCS1 and LCS2 classes, as well as the catamaran HSSV, all utilise aluminium FSW.

Civilian vessels too utilise FSW, often for the modular build of aluminium decks and superstructure components on a steel hull to provide strong, stiff but lightweight upperworks.

3. TECHNOLOGY TRANSFER TO STEEL

3.1 FSW TOOLS

The greatest obstacle to transferring the technology of friction stir welding from light metals into steel has been that of finding a tool material with which to stir the steel. Aluminium starts to soften at 300°C and flows sufficiently well for friction stir welding at 450°C. FSW tools made of steel are able to withstand such temperatures and retain adequate strength, and also tend not to react with most types of aluminium with the exception of 7xxx grades. For welding these higher strength grades, and aluminium thicker than around 40mm, the FSW tools are typically made of nickel cobalt superalloys such as MP159. Steel, however, does not soften and flow sufficiently for friction stir welding until almost twice the temperatures experienced with aluminium, typically around 900 to 1,000°C. Few materials retain adequate strength, toughness and abrasion resistance to stir steel at such temperatures and of those that do, the problems of steel's high chemical reactivity results in rapid tool wear as the hot steel effectively alloys with and dissolves the tool.

Efforts to FSW steel with tools based on the refractory metal tungsten generally failed due to the very high wear rates experienced, the tools rarely achieving more than a few metres of weld in useful thicknesses of carbon steel before they wore to the point that they ceased to be able to weld. Tools based in ceramics were also tried, these having the advantages of good wear resistance, adequate thermal resistance and acceptable chemical inertness to hot steel, but these were unable to withstand the forces and bending stresses required to stir the steel. [3,4]

Had it not been for the extremely good mechanical properties observed in the short welds that were able to be made in steel, under laboratory conditions, then the process would probably have remained limited to use in the lower strength, easily plasticised metals. However, the potential to obtain the same benefits in steel as had been seen in aluminium, magnesium and copper acted as significant drivers for further development and FSW tools are now being introduced into the market that can make industrially useful lengths of weld in steel.

Several tool manufacturers, most recently MegaStir Inc of the USA and Element Six in the UK, have developed FSW tools for steel based upon a hybrid materials technology, combining hard wearing and chemically resistant superhard ceramics with a strong, tough, metallic component in order to produce a tool capable of welding industrially useful lengths of steel under industrial rather than laboratory conditions. [5]

3.2 CURRENT CAPABILITY

Element Six's latest FSW tools for welding 6mm thick steel have been tested at TWI and shown to consistent from batch to batch, and capable of producing defect free welds in carbon steel under several different testing conditions. These test conditions were:

Regimen 1: Multiple 2m welds

- Multiple tool plunges into cold, hard steel;
- Tool operating several minutes at elevated temperatures;
- Typical of pipeline welding, repair welds, assembly work.

Regimen 2: Multiple 5m welds

- Fewer tool plunges into cold, hard steel;
- Tool operating typically 20 minutes at elevated temperature;
- Typical of modular construction, pressure vessels.

Regimen 3: Fewer, but longer 20m welds

- Two or three tool plunges into cold, hard steel;
- Tool operating at elevated temperature for an hour or more;
- Typical of panel production for ships.

The welds were made at a welding speed of 300mm/min in S355J2+N steel. Under all service conditions, the tools were capable of producing consistent, defect free welds. All tools tested reached an accumulated weld length of 40m, with no failures experienced during the trials. Testing is still ongoing to determine the ultimate longevity of the tools.

The high strength and elevated temperatures required to friction stir weld steel currently impose thickness limitations on the process. This is due to limitations inherent in the tool materials rather than the FSW process itself: when welding aluminium alloys plates of up to 75mm thickness have been welded in a single square butt pass and up to 160mm has been reported in a two pass weld. The tools currently entering the market for steel are aimed at welding steel from 2 to 6mm in thickness, with a second family of tools under development for thicknesses up to 12mm. Using a two pass welding process, this would allow steel up to around 20mm to be welded if access was available to both sides of the joint line. Longer term, further development of the tool material may allow even greater thicknesses of steel to be welded with several potential users having already expressed a desire to make welds in high strength, high alloy steels up to 40mm thick. Under laboratory conditions, prototype tools from MegaStir Inc have produced defect free welds in 19mm thick A752-65 steel plate. [6]

4 WELD PROPERTIES

4.1 STRENGTH

The fine grained microstructure seen in friction stir welds generally provides them with tensile strength properties closer to the parent metal than is typically the case with conventional fusion welding. With steel, this benefit is further enhanced by the fact that welding takes place in the transformation temperature range and careful selection of the welding parameters can exert a useful degree of control over the phase transformations that take place during the welding process. It is possible, for example to make welds optimised for strength or for toughness, or a combination of both, depending upon the service requirement. It is generally the case that friction stir welds in steel are found to be stronger than the parent metal in which they were made. Where friction stir welds are made between dissimilar grades of steel, for example a carbon steel and a stainless steel, the failure tends to occur in the weaker of the two parent metals and away from the weld zone.[7]

A number of samples were taken from a single pass butt weld made in 6mm thick S355 steel with E6 tools that had accumulated more than 60m of welding and subjected to mechanical testing. These samples passed a root and face bend test, an indication that the samples had no significant defects and had a good degree of ductility in the weld zone. The cross weld tensile test specimens failed in a ductile manner in the parent metal some distance from the weld and HAZ. The samples are shown in Figure 8.

The mean ultimate tensile strength of the tested samples was 562 MPa. No parent metal was tested but EN 10025-3 for S355 steel in the fine grained, rolled condition specifies a minimum UTS requirement of 520 MPa for steel less than 40mm in thickness. As the welded samples failed in the parent metal it can only be stated that the weld strength was at least equal to 562 MPa and exceeded the minimum parent metal requirement.

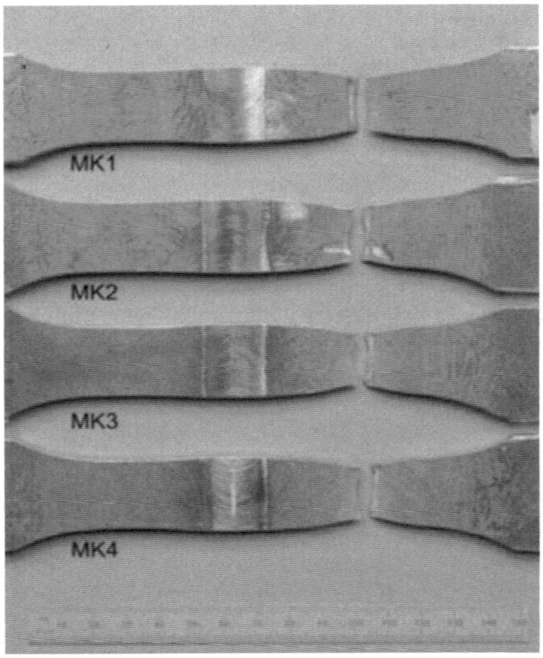

Figure 7. Cross weld tensile test samples of FSW in 6mm thick S355 steel. The samples show ductile failure in the parent metal away from the weld zone.

4.2 TOUGHNESS

The fine, multiphase microstructures generated by steel FSW give rise to welds that can also be tougher than the parent metal. Samples were taken from a FSW weld in 6mm thick S355 steel produced by a tool that had made over 60m of accumulated weld and subjected to Charpy impact testing. The Charpy impact testing was performed at -20°C and in accordance with BS EN ISO 148-1:2016. The samples were removed from mid weld thickness and notched in the welding direction through the thickness on the weld metal centreline.

Impact strengths of 49, 57 and 61 Joules KV were recorded for the three test pieces, with the mean being 56J. This is considerably in excess of the minimum specified parent metal value for S355 at -20°C, which is 27 J KV.

4.3 FATIGUE

The complex, mixed microstructure that can be generated in friction stir welds appears to be capable of providing excellent fatigue performance. Work to study this effect is ongoing but it appears that the welds are remarkably resistant to crack propagation, even in the presence of sharp edged cracks. Figure 9 shows a metallographic section through a test piece extracted from a friction stir welded , 6mm thick plate of DH36 steel. Two deliberate defects were incorporated into the weld: a lack of root penetration and a lack of sidewall fusion at the weld cap. When subjected to both simple tensile testing, and fatigue testing, the samples failed in the parent metal and not in the weld.

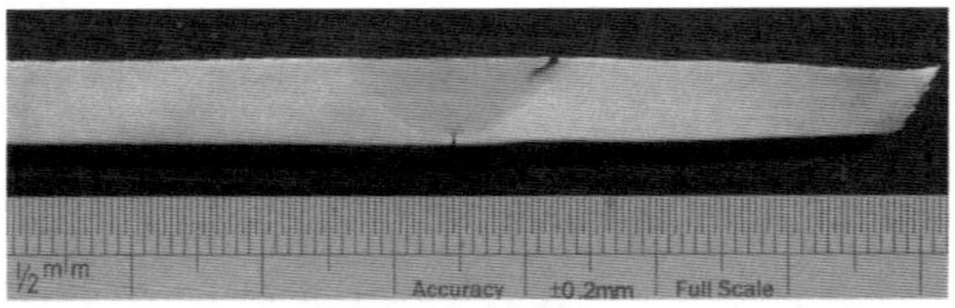

Figure 8. A metallographic section through a deliberately defective FSW in 6mm thick DH36 steel. The failure site is in the parent metal to the right hand side of the sample.

4.4 DISTORTION

A major benefit of the FSW process is that it produces very little distortion. As the metal being welded does not melt, and thus does not pass through a solid to liquid to solid phase change with its associated volumetric change (up to 4% in the case of some alloys), the distortion induced in the welded fabrication is significantly less than seen with conventional arc or laser welding techniques. The heat input to the weld zone in FSW is far more localised, and generally more rapidly dissipated, and the work pieces rigidly clamped, thus further reducing the potential for distortion. [8]

Figure 10a and 10b show the distortion measured in identical test coupons manufactured by Submerged Arc Welding (10a) and FSW (10b). The SAW test piece, made in 6mm thick DH36 steel using state of the art equipment and experienced welders, had a maximum distortion of 110mm along its length. The first test piece manufactured by FSW exhibited a maximum distortion of 15mm, which was subsequently reduced to just over

2mm on a second attempt with revised welding parameters.

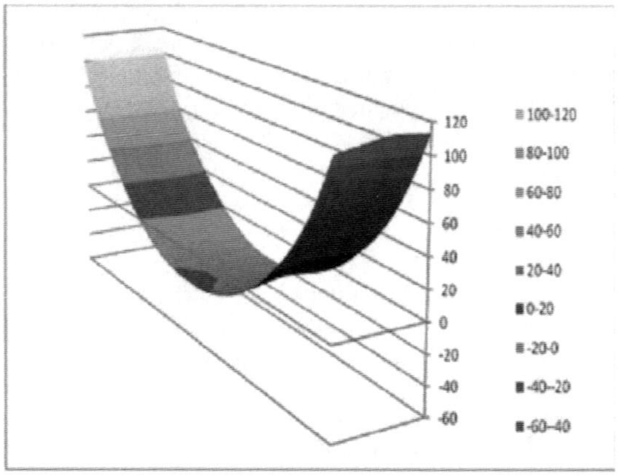

Figure 10a (above) showing the distortion measured in a SAW and 10b (below) showing the distortion in a FSW, both in 6mm thick DH36 steel. (Data courtesy of BAE Surface Fleet)

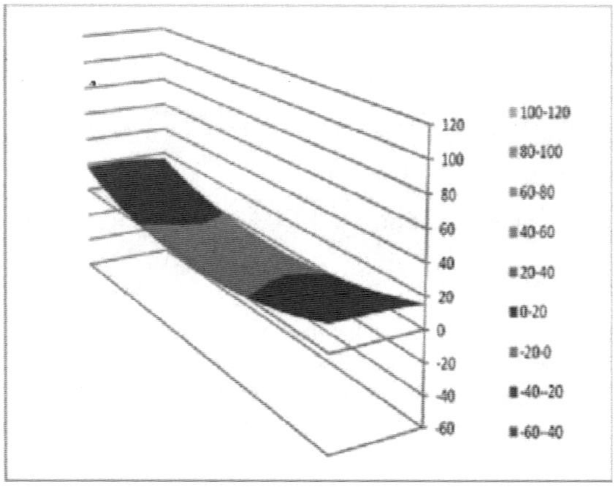

4.5 DISSIMILAR WELDS

As FSW is a solid state, i.e. no melting, welding process it is far less influenced by the alloy content of the steels being welded. It is therefore possible to weld different grades of steel together more easily than with other techniques, including carbon to Duplex stainless steels, and to weld steel grades that are found challenging to weld by other processes [9].

Figure 11 shows a dissimilar metal weld made between S355 carbon steel and S32205 Duplex stainless steel. From the micrograph it can be seen that no alloying has taken place between the two different steel grades, the joint is made purely by mechanical mixing. No alloying elements have been lost from either of the two grades in the weld zone, nor has elemental segregation occurred in this region. The Duplex stainless steel retained the phase balance of the original parent metal, albeit with a much finer grain size in the weld zone.

Samples taken from the weld passed both face and root bend tests, and under tensile testing the sample failed in the parent carbon steel away from the weld area.

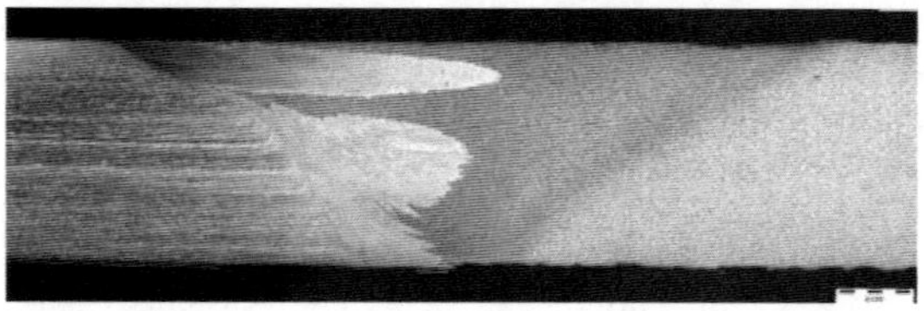

Figure 11. Metallographic section through a weld made between 8mm thick S355 carbon steel and S32205 Duplex stainless steel.

Interest has been shown by the shipbuilding industry in the potential of the FSW process to weld metals with very dissimilar plasticisation temperatures, for example joining aluminium to steel. Though this is difficult, with aluminium for example having a melting point considerably below the softening point of steel, aluminium to steel welds have been achieved at TWI. Samples taken from the fabricated plate for tensile testing failed in the parent aluminium rather than the weld, indicating that the weld strength was at least as good as that of the weaker of the two parent plates. Further work would need to be done to investigate the long term properties, particularly fatigue and corrosion of such dissimilar metal joints before it was adopted for industrial use.

5. PROJECT RESURGAM

Given the availability of effective and increasingly reliable tools for steel FSW, TWI established a consortium of companies in the shipbuilding sector to bid for EU funding to identify investigate potential applications for steel FSW, resulting in the RESURGAM project. [10]

The RESURGAM (Robotic Survey, Repair & Agile Manufacture) project will combine the development of FSW technology for steel with the introduction of digital manufacturing techniques for both the initial modular fabrication of ships across multiple sites and the in water repair of damaged ships or marine structures.

The FSW element of the project will look at developing a suite of FSW tools able to weld multiple grades of steel from 2 to 12mm in thickness, in air and under water. These tool developments will be accompanied by the development of a modular FSW system that can be retrofitted to existing CNC milling and machining centres and a robotic system that can be deployed underwater to make *in situ* repairs [11]

These fabrication and repair capabilities, backed by the secure, digital Industry 4.0 infrastructure and techniques already in widespread use in the automotive and aerospace industries, will facilitate the rapid, coordinated but distributed modular manufacture of ships and watercraft throughout Europe.

Practically, this will allow ships damaged anywhere in the world will have the option of being repaired in place without the need to travel to the nearest dry dock. This will allow ship owners to choose the most suitable yards to conduct their repairs rather than the nearest, and the repairs may be undertaken by yards with no dry dock of their own thus significantly increasing the number of yards able to undertake such work.

6. CONCLUSIONS

FSW has a proven history in the marine sector for the fabrication of small to medium sized aluminium ships. Innovation continues in further developing the process for aluminium fabrication, particular for corner and fillet welding, but the recent development of tools from several manufacturers that are capable of welding steel now opens up the prospect of applying the process to the fabrication and repair of larger steel ships and potentially, offshore structures such as oil production platforms, pipelines and wind turbine towers.

Considerable further work remains to be done to deploy steel FSW industrially:

- technically to develop the tools and welding process;
- with potential users of the process to tailor it to industry's needs;
- and with the classification societies to ensure it meets their acceptance.

Project RESURGAM provides an immediate route forward for the technical development programme, and TWI is already working with classification societies such as Lloyd's Register to address the requirements of the Certification process. The involvement of naval architects, designers and potential users of the process is now sought to ensure that the process meets the needs of industry.

7. ACKNOWLEDGEMENTS

The RESURGAM project has received funding from the European Union's Horizon 2020 research and innovation programme under grant agreement No 101007005

8. REFERENCES

[1] "FSWBOT: Weld repair of internal corrosion defects in live pipelines," S Cater, C Stanhope, J Sykes, P Routeledge and P Wilson, OPT 2020, Amsterdam, 2020.

[2] "Friction Stir Welding of Steel", S Cater and J Perrett, TWI Members' Report MR1008, 2011

[3] "Evaluation of Tool Material Performance for the Friction Stir Welding of 6mm AISI 304L Stainless Steel", J Perrett, TWI Members' Report MR915, 2008

[4] "Friction Stir Welding of Steel: Process considerations", S Cater, TWI Members' Report MR1035, 2013

[5] "Advanced tool material technology for FSW stainless steel", J Perret, TWI Members' Report MR951, 2010

[6] https://megastir.com/the-right-tools-for-the-job/

[7] 'Development of a Process Envelope for Friction Stir Welding of DH36 Steel – A Step Change'. Toumpis A, Galloway A, Cater S and McPherson N, 2014: Materials & Design, 62, 64-75.

[8] 'A comparison between single sided friction stir welded and submerged arc welded DH36 steel plate'. McPherson N A, Galloway A M, Wood J and Cater S R, 2012: 9th Trends in Welding Research Conference, Chicago, USA, June 2012.

[9] 'Friction stir welding of PM2000 alloy'. Dawson K, Cater S, Tatlock G J and Stanhope C, 2014: Materials Science and Technology.

[10] https://cordis.europa.eu/project/id/101007005

[11] https://www.e6.com/en/about/News/ELEMENT-SIX-PCBN-TOOLS-FSW-EU-RESURGAM-PROGRAMME

9. AUTHORS BIOGRAPHY

Stephen Cater, FRIN, MEI, AssocRINA, AWeldI, holds the current position of Principal Project Leader at TWI Ltd. He is responsible for managing the friction stir welding element of the RESURGAM project. His previous experience includes the development of the FSW process for steel and for thick section (>40mm) aluminium, as well as developing FSW for use under water and in oil. A metallurgist by training, and winner of the Richard Weck Award for his work on steel, Stephen has published over 20 papers on FSW.

GENERATION OF THE DIGITAL TWIN OF A PROCESS PLANT BASED ON DEEP LEARNING

M Grau and **J Stjepandić,** PROSTEP AG, Germany

SUMMARY

Nowadays, Digital Twin is used both as a subject of scientific research as well as a buzzword to attract existing commercial products and services. It poses one of key innovations in the IT related products and services in the past years. There are numerous types of Digital Twin, in particular in the manufacturing domain. Big challenge still remains how to generate the Digital Twin for existing objects in a plant. In this paper, an approach for the generation of Digital Twin in a built environment based on scan and object recognition is presented which is commercialized as an extension of a well-running portal OpenDESC.com. After the theoretical foundation and the practical requirements are described, the functional approach and the practical implementation are explored. The examples from manufacturing plant and process plant are presented. The method for assessment of results is discussed. Finally, conclusions and outlook summarize the insights and give future directions.

NOMENCLATURE

[Symbol]	[Definition] [(Unit)]
AI	Artificial Intelligence
DT	Digital Twin
ML	Machine Learning
PLM	Product Lifecycle Management

1. INTRODUCTION

Over the last few years, a concept called Digital Twin has raised rapidly as a comprehensive approach in the industrial domains. As a recently coined buzzword without a clear definition, it comprises a description of a deep coherence between the physical products, processes or services "as delivered" and their virtual surrogates [1]. Digital Twin is used also in process industry. Like other assets of a process company, the modern plant is also planned and documented using powerful software tools which maintain high-fidelity plant models. A digital plant layout is the main digital representation of a plant ("Digital Master" as a type of Digital Twin) which includes all necessary constituents and equipment (building, machinery, and auxiliary equipment) [2]. In the focus of the layout planning are the design and the maintenance of the process plant which determines the productivity and feasibility of the entire plant [3]. To achieve an optimal planning result, it is necessary to share ideas between several teams or employees and functions which can be made using collaboration front-end of planning system [4]. Virtual plant layout is beneficial for communication and knowledge transfer. Furthermore, it is often required by regulatory bodies [5]. In case of a new project, physical plant is built based on the virtual plant. In the opposite case, if a built environment must be modifies based on poor or not existing 3D documentation, the virtual layout must be recreated ("Remastering"). One of the most frequent questions in the field of digital transformation is how to handle the physical objects which do not exist as an appropriate digital model. Such objects include complex long-living spatial objects like plants, machines and industrial equipment, in particular which are produced once and used in operation and maintenance in a long period. This process can be done either fully manually (measurement and creation of CAD models), partially manually based on scans or almost automatically (based on scans and object recognition). The scope of this paper is the method for generation of the Digital Twin of a process plant based on scans and object recognition using Deep Learning methods. The remainder of this paper is structured as follows: in section 2 the background is given. In section 3 object recognition methods are presented, following with the practical approach in section 4. Results are discussed in section 5, following with conclusion and outlook in section 6.

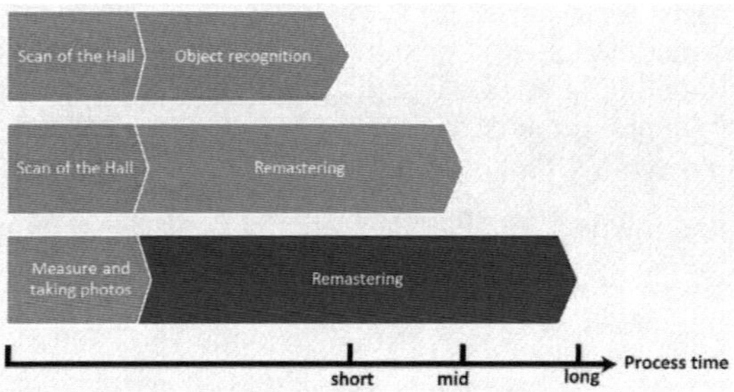

Figure 1: Approaches to create a Digital Master [6]

2. BACKGROUND

Driven by a tremendous development in areas related to computer vision, a plethora of object recognition approaches and methods were developed recently for different purposes. Although photogrammetry still plays an important role, a trend to point cloud-based approaches is obvious which is the basis for object recognition algorithms. For industrial application with rough environmental conditions (dirt, smoke, darkness, dust, vibration), it looks like the best compromise between costs, accuracy, acquisition speed and robustness). A scanner can be used as a terrestrial and mobile device (carried by human, car, drone) [7]. To prepare the point cloud for a feature-based process of planning and simulation, the contained objects must first be recognized and a CAD-based structure must be created. The great challenge in every process chain is to implement the desired function with minimal effort and losses. As one of the alternatives, the object recognition is selected to implement the layout planning of a plant in 3D. Here, the challenge is how to embed the object recognition in the process of virtual 3D layout planning in a built environment as well as which findings and results can be expected. Realistic 3D layout models need to be created using point clouds acquired by commercial terrestrial laser scanner and prepared for object recognition with methods of Artificial Intelligence (AI) considering the strict data quality requirements. Seamless, robust, (semi-) automatic workflow of primarily standard, modular components with low user assistance is of particular interest in order to achieve high efficiency. A proper quality assessment and error handling is necessary to facilitate automation of entire process [8].

3. OBJECT RECOGNITION METHODOLOGY

Object recognition basically comprises multiple objects of different categories in a scene (e. g. plant hall). The generic process of the object recognition is shown in Figure 2. At first, an object must be acquired by an observation which can be done by devices through various physical principles (photo, video, distance measurement) as described in previous chapter. A commercial, productive technology was used in this project. In the next step, the features which can be identified in the scene must be classified against a patterns database. The recognized features are integrated in the next step to objects by using external features. By using environment knowledge, the result of this process is the recognized object, based on a recognition accuracy which must fulfil the requirements of this specific process chain [9].

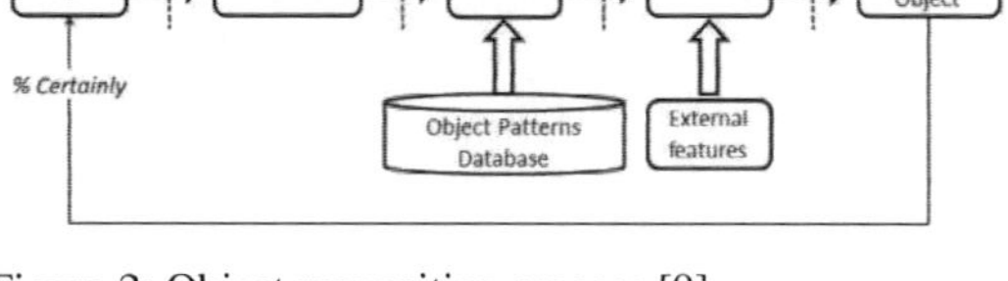

Figure 2: Object recognition process [9]

Objects in a plant have a few distinct external characteristics: some of them are enclosed by a cover (e. g. container), the others (e. g. pipelines) discover a distinct dimension. Even if objects in the plant often touch each other, there is rarely an overlap, but the visual occlusion is very common. It must be compensated by an appropriate acquisition rule. From a geometrical point of view, objects consist almost exclusively of standard geometries with many prismatic parts. This opens the opportunity to combine the steps of classification and integration in Figure 2 by searching for and recognizing the entire objects rather than the individual features which would be collected to an object further. This assumes that the external shape of an object is distinctive enough and the object pattern database large enough to ensure that the object can be always recognized [10].

Embedded in a working environment, the practical procedure can appear as follows: instead of identifying the detail feature of each object, the identification of the corresponding CAD models from a reference library and the transfer of geometry and other object data (e. g. valve types) as modular objects directly from the library [11] to the design context can significantly reduce the scan times for a first rough pre-scan of the plant. At the same time, database reconciliation enables the use of simpler and cheaper search methods [12]. After a specific object has been recognized in the right position, its native CAD model can be used in the downstream processes, e. g. plant re-design to generate Digital Twin.

Methodology for object recognition can be derived according to various criteria. The most important criterion is the input data type. This can be an image, a video, or a point cloud of an object or a scene. A further subdivision can be made for certain input data type. This includes the object representation and the type of experiment how the sample results are achieved. A large test base is necessary for both the detection of the progress in development of an object recognition method and comparison of results achieved by different object recognition methods. The ideal goal of generic object recognition is to develop general-purpose object recognition algorithms achieving two competing goals: high quality and accuracy as well as high efficiency [13]. High quality recognition must accurately localize and recognize objects in images, point clouds, or video frames, such that the large variety of object categories in the real world can be distinguished (i. e., high distinctiveness), and that object instances from the same category, subject to intraclass appearance variations, can be localized and recognized (i. e., high robustness). High efficiency requires the entire detection task to run at a sufficiently high frame rate with acceptable memory and storage usage. Focus lies here on object recognition based on point clouds generated by a scanner [14].

A critical aspect in the procedure for object recognition is the proper test base. To train a deep 3D shape representation with a significant intra-class variance requires a large collection of 3D shapes. Since this requirement usually appears in every object recognition project, there are a limited number of datasets that can be used to train and test 3D object classification methods for indoor and outdoor applications. The most popular data set at this time is ModelNet with over 150.000 samples [15]. No international standard is known on this area at this time. Many CAD datasets, created in the past decade, were limited both in the variety of categories and the number of examples per category. These data sets can be used to train the classifier.

The rough point clouds of scenes in a plant, gained either by scans or from images and videos with subsequent dense point generation, cannot be used directly for the object recognition objects for several reasons. Such a point cloud contains false or undesired information, e. g. surroundings or reflections by large glass surfaces. Therefore, the point cloud needs to be prepared for seamless further processing, e. g. object recognition.

Four transforming steps are needed to make a point clouds of a scene meet the application requirements: point cloud preprocessing, mesh reconstruction, point cloud segmentation and point cloud modelling [16].

The aims of the point cloud preprocessing are to prepare the point cloud gained by a device (scan, camera) for further processing in several transformation and translation steps by provision of point clouds with high quality for downstream operation. Generally, the step contains four sub-steps, i. e. registration, noise filtering, outlier removal and down-sampling. Which of the four sub-steps are needed depends on the application requirements and the quality of the point clouds. For example, if the application requirement is a combined point cloud and the existing point clouds are separate, then registration is required which merges two point clouds into a consistent point model by a spatial transformation.

The aim of the mesh reconstruction is to generate a mesh model of the object of interest by using the point cloud obtained from previous step in cases when the object recognition does not directly process the point could. Typically, the algorithm used for mesh reconstruction is Poisson surface reconstruction (PSR), which reconstruct a watertight, triangulated approximation to the surface of the object according to its point cloud.

The aim of the point cloud segmentation is to segment a point cloud and obtain the points of the objects of interest. Point clouds are classified into multiple homogeneous areas, the points in the same region will have the same properties. The segmentation is challenging because of high redundancy, uneven sampling density, and lack of explicit structure of point cloud data.

The aim of the point cloud modeling is to generate a geometric model of the object of interest by using the point cloud segment obtained from the previous step. Basically, there are four ways to create a model (Figure 3) [17]:

- Identification of an annotated model (or object) and its position,
- Recognition of a non-annotated model (or object) from a model library and its position,
- Recognition of the singular features of a model and the composition of the model itself by recognition of several features in the right position, and
- The recognition of "dumb" objects by manual remastering of a model using a point cloud segment.
- The final composition of pipeline systems which contains the semantics.

It is worth to note that some portions of point cloud remain not recognizable.

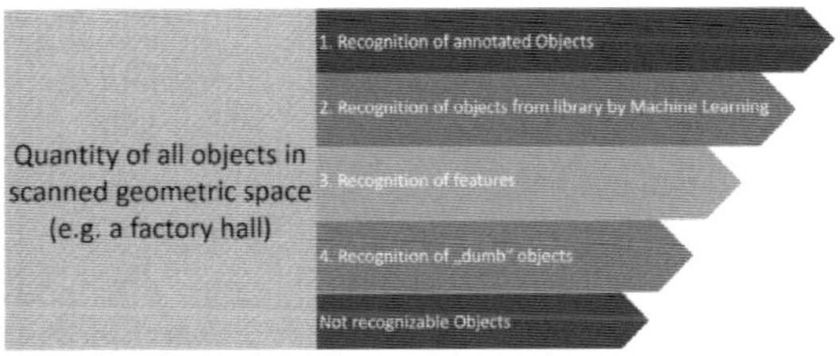

Figure 3: Alternatives for model creation

To generate a geometric model (variants 3 and 4) of the object of interest, it is essential to obtain the model parameters based on the results of point cloud segmentation. Generally, the method used for obtaining model parameters is obtaining dimensions of objects and applying the modelling functions of CAD systems. In contrast to models consisting of features (e. g. pipeline), several objects have not a fix form (e. g. flexible pipelines) and, thus, must be explicitly modelled as "dumb" models.

The geometric model of a pipeline system is handled as a collection of cylinder segments with connectors and thus the radius is one of its model parameters. The radius can be obtained by computing the shortest distance from the centerline to the point cloud of the pipeline obtained from point cloud segmentation, or by projecting the point cloud of the pipeline onto a plane perpendicular to the centerline of it to generating a resulting circle and computing the radius of the circle [10]. With the model parameters, a geometric model of object of interest can be easily generated.

For each recognition operation, the recently developed Deep Learning techniques based on point clouds are used. This trend is heavily supported by development of lidar devices, as a cheap, efficient, and rapid remote sensing technique. Deep learning, through an end-to-end model, directly performs a nonlinear transformation layer by layer on the original input from a lower level to a higher level. Hence, it can automatically extract the information contained in massive data, which greatly simplifies the learning process [17].

At the end of the modelling phase, the reconstruction of semantics must be done, for example, to connect pipe segments with valves and vessels to a pipeline.

4. ENTIRE PROCESS

4.1 AUTOMATION OF WORKFLOW

Complex procedures as described in previous section usually are being controlled by appropriate workflow software. Prior of such an implementation, it must be defined how the process steps scan, object recognition and generation of Digital Twin (Figure 4) interact with each other [18]. Basically, these are tasks for different user profiles because the present technology does not provide the performance to collect entire functionality for generation of Digital Twin in one device (e. g. portable computer). The tasks scan, object recognition, design engineering (to provide the requested change) and process planning (generation of Digital Twin) can be seen functionally independently of each other; with data exchange among singular steps [19].

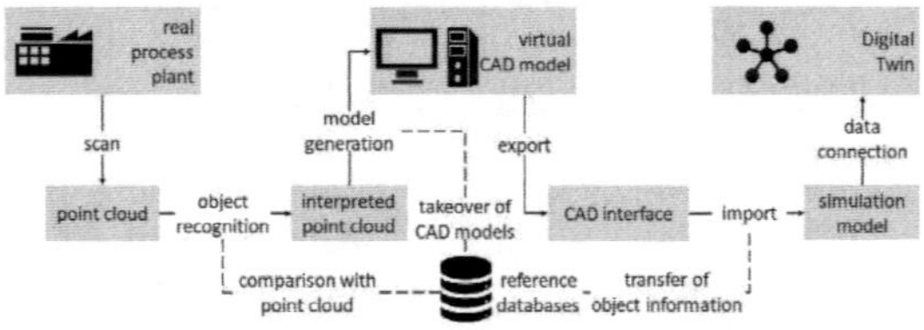

Figure 4: Workflow generation of Digital Twin

Assuming a request for reconstruction, overhaul, or maintenance, the object acquisition must be provided at the plant site by any device (scanner or camera, terrestrial or mobile) and the raw data are transferred by a secure web connection or disk to the next instance which provides data processing (object recognition, feature recognition, translation) as described in the previous section. After the full data model of the plant was created, it is sent to the design and engineering which complete the required change of the process system. Finally, the data set is being exported to the process planning department to conclude simulation, adaption and optimization.

4.2 MANUAL ASSISTANCE

While some steps must be done manually (point cloud preparation, noise filtering, outlier removal, feature extraction, and almost the entire training), the workflow must ensure an automated execution of all remaining steps. This is in particular true for the most time-consuming steps: execution of segmentation and execution of object recognition. It also includes the handling of alternative routes and loops necessary to execute all the options to create a feature model. The flexible, modular handling of the object recognition framework with the option of replacement by an another must also be provided [20].

Further important aspect is the level of automation of singular steps. While singular data sets differ from each other, for sake of efficiency is it necessary to estimate the share of manual assistance in the entire process. The device-oriented tasks already are highly automated and can provide a dense point cloud as input for a high-performance point-cloud editor, in particular if a scanner is used.

The most remaining manual work can be expected on the segmentation of complex scenes which promises the highest rationalization potential, although no support from publicly available directories can be expected. A successful segmentation anticipated a reduction in data volume of point clouds with a factor of 1.000 and more between the rough point cloud and the segmented point cloud focused on segment for further processing can be achieved. This drastically reduces the demand on hardware, the processing time and makes the interactive work much easier.

4.3 MODEL PREPARATION

Each point cloud must be prepared for the object recognition. At first, the sighting must be premised to visually check the output of the data acquisition procedure. It is possible that the automatic registration has failed. Therefore, the data set must be positioned in space. It occurs by selection of an appropriate reference coordinating system. During these manual steps, the model clean-up, the noise filtering and the outlier removal are conducted. Eventually, the entire point cloud must be splitted, e. g. to extract the portion of interest which will be processed further. The extraction of models and portions for training of recognition respectively segmentation occurs at this point of time. Finally, the point cloud must be exported in the desired format (e. g. e57).

From this point on, the process runs automatically apart of training activities which are not mandatory for each data set. Training is necessary in a sufficient extent to provide the base for the later testing [21]. Mesh reconstruction and point cloud segmentation are implemented as separate utilities for the object recognition framework. It is necessary for voxel- or surface-oriented recognition frameworks. To collect all automatic running steps, a user front-end was built (Figure 5). It controls the semantic segmentation, cluster extraction and cluster classification (recognition of singular objects). This tool also includes a viewer allowing the visualization of each intermediate result.

Although good object recognition algorithms are capable to search for objects in each space, for an intended industrial exploitation some pre-processing and adjustment in sense of subdivision (segmentation) of huge spaces (and data volumes vice versa) are beneficial for several practical reasons. At first, it helps to generate a structure of a scene which aims to identify specific points of interest. Secondly, it can help to extract unnecessary data (remaining outliers, objects which were scanned by mistake) in a structured manner. Thirdly, the segmentation helps facilitates adding of further information (e. g. buffer, hold places, restricted areas). Fourthly, it is much easier to validate the object recognition results and handle the failures and errors in a restricted, smaller space. Finally, a segmented scene allows the use of different recognition methods for different classes of objects (e. g. vessels vs. pipelines).

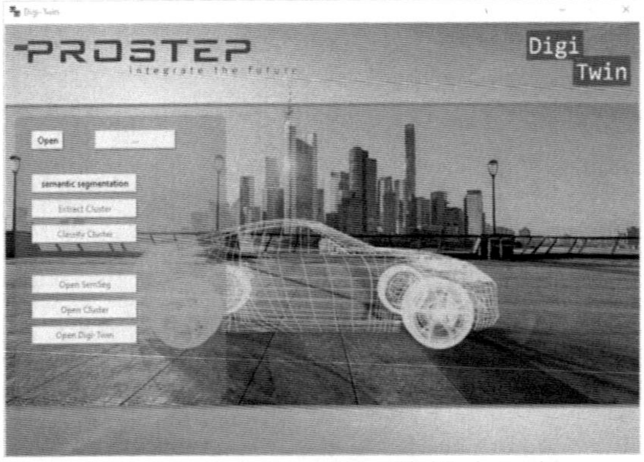

Figure 5: The user interface of the DigiTwin tool

Segmentation can be conducted manually or by supervised training which is the preferred way for complex scenes. Therefore, additional methods were developed to simplify object recognition preserving the data quality. The first issue is the preselection of smaller agglomerations in the entire point cloud which are assumed to contain an object. Such clusters can be covered by a bounding box like a packing box for a product in the right position and orientation. It helps in the assessment of results too. The segmentation module is being trained to subdivide the entire scene in nine categories: roof, wall, floor, pipelines, structure, vessels, transport, auxiliary, and noise. For a process pant, pipelines and connected devices are of a particular interest.

4.4 TRAINING AND TESTING

Before the object recognition can start, the objects of interest must be available in the system library. Since they are usually not included in the standard database such as ModelNet, an additional reference library must be set up that contains the objects of interest. This is performed in two ways: either by using CAD models (e.g., vendor library) or exemplary point cloud models. To ensure the similar properties of both model types, CAD models are transformed in point clouds using a virtual scanner which is stochastically moved around the object.

Singular point clouds models must be further transformed to prepare an adequate input for object recognition procedures, depending on the object recognition framework used [22]. In case of the voxel-oriented framework, the point cloud must be translated into a corresponding voxel model. The side length of a voxel plays a significant role here. Tests have shown that a variable length which is related to the object shape provides the best results. Although the start model already has been simplified, a further simplification occurs during the translation to a voxel model which hardly can be prevented. If the simplification dilutes a local feature (e. g. an auxiliary device), then sufficient recognition results cannot be expected. Rather, it is to be feared that a wrong object will be recognized. Therefore, when building the sample models, care must be taken to ensure that the distinctive features are considered.

An exemplary procedure in three steps (segmentation, clustering, recognition of object in singular clusters) has been trained with up to 1000 epochs, the developed framework can be tested using real-life data extracted from the process plant.

5. RESULTS

The new DigiTwin tool was tested using the input from two industrial use cases: (a) a manufacturing plant with 4 halls, and (b) a biogas plant located in north of Germany. In both cases, the object recognition of objects available in the library was of the particular importance. The models were manually prepared and put in the tool chain as described in the previous section. The overall recognition rate is high, although the environmental impact was quite negative. More than 75 percent of all objects can be truly recognized. However, the recognition procedure is sensitive and needs to be stabilized. Provision of bounding boxes for singular object of interest is helpful for repair of missing and wrong oriented objects.

6.1 MANUFACTURING PLANT

The assembly in the CAD system SolidWorks represents an industry standard in the field of industrial engineering and planning. The hall is divided into several cells, each containing machines and devices, which can then be further subdivided as required. Apart of the training process which lasts two days on a desktop computer, this automated process takes few minutes of computing time with potential for further improvement. For further work (e. g. customization design) a complete, consistent CAD model is necessary, which is validated by a quality check [23]. If some components are missing here, they must be added later. Missing geometry can be created and adapted by using standard CAD functions. The bounding box is of great advantage here because the missing component is placed in it. Once the assembly has been completed (Figure 6), it can either be used in a customization design or exported standard CAD format for simulation tasks which can be provided by a powerful simulation software [24]. Ultimately, this is the highly automated procedure how Digital Manufacturing Twin is created in a built environment [25].

Figure 6: Results of the object recognition in a manufacturing plant

6.1 BIOGAS PLANT

To verify this approach for the recognition of pipelines, further experiments with the point cloud from the plant construction (biogas plant) have been conducted. This method is only based on the registered industrial point cloud itself regardless of the varying point densities [26]. Such industrial point clouds contain defects such as occlusions and sparseness. The results were disparate. Good results were achieved in the segmentation phase automatically and robustly. Pipeline easily can be separated from the remaining components of a plant. In highly dense industrial spaces, class segmentation of cylinders, I-beams and valves that have easily distinguishable geometric patterns in the processing unit, warehouse and the oil refinery work well. In a first attempt, the object recognition did not perform sufficiently. There are several reasons for this. Pipelines cannot be viewed as unique objects. Therefore, the additional object recognition algorithm was necessary.

The focus of a biogas plant lies in the recognition of the pipelines including their semantics (Figures 7 – 11). For this task, a specific extension of the object recognition framework was developed. The step segmentation runs as previously described. The step clustering consists of the identification of the singular pipe segments which are subsequently added to pipelines. By detection of the corresponding pipe radius, the centerline of the pipeline can be derived.

Figure 7: Point cloud model of a biogas plant

Figure 7 shows the point cloud model of the entire biogas plant which was obtained by scans from different devices.

Figure 8 shows the result of segmentation. Good to see the distinction between ground (green), walls (light blue), roof (dark blue), ventilation (violet), vessels (purple) and pipelines (yellow). This model is still a point cloud model, but with a drastically reduced data volume.

Figure 9 shows the extracted pipelines which would be subject of redesign and optimization. Worth to note that the entire data volume was reduced from 57 GB for the entire plant to only 22 MB for the pipeline.

Figure 8: The segmented point cloud model of a biogas plant

Figure 9: Point cloud model of the pipelines

This intermediate result provides a good basis for design work on reconstruction and optimization because leading CAD systems can handle a smaller amount of point cloud models.

Figure 10: Resulting model of the pipeline with the centerline and radius

Figure 11: Resulting model of the pipeline with the centerline and radius in context of the entire plant

Finally, Figures 10 and 11 show the resulting pipeline alone and in the context of the entire plant, respectively. Good to see the centerline (red in both figures) and the pipe diameter in Figure 8. This result is obtained in a neutral format and can be exported to each proprietary CAD format for further design work.

6. DISCUSSION

Assuming that a searched object is known, either in the publicly available or in the private library, there is a high probability that it will be recognized in the correct position and orientation. The distance deviation is acceptable for this purpose it remains in the region known from literature [27]. Manual rework in a small extent can be always expected, indeed.

Several improvements are planned for the next development steps:

- The huge amount of input data could be significantly reduced by a better scanning procedure, if architectural domain knowledge can be used for this purpose. That could be achieved by a pre-scan of objects of interest as known from autonomous driving.
- By using low-cost devices (e. g. augmented reality glasses, 360-degree camera), the usability could be improved, and the costs reduced. This would help to prevent the impact of occlusion. However, the concern of the low position accuracy still remains.
- The final result could be improved by extension of the object library, for example by a blend with further libraries (CAD standard part libraries). Industry circles and associations could provide their data.
- While the training requires so much effort, its automation would drastically reduce the processing time and increase the overall efficiency.
- Integration of PI&D and non-geometric information would improve the confidence of the regulatory bodies.
- While the object recognition basically is a probability calculation, it needs an extensive and quantitative comparative performance evaluation in order to meet the requirements from the industry.

7. CONCLUSIONS

The legitimacy of object reconstruction with semantic enrichment needs to be primarily justified in terms of cost-benefit [28]. If the object recognition as shown here in two use cases from the industry is successful, the justification is easy. However, the drawbacks must be considered to prevent a use case with no recognition (e. g., on significant occlusion which cannot be compensated). Theoretically, in highly dense industrial spaces data acquisition could be made impossible both by a missing accessibility though the acquisition device and the poor recognition caused by incomplete point cloud and occlusion [29]. Therefore, such obstacles can only be resolved by an incremental scanning (e. g. by a highly flexible mobile device which can access each segment of a plant) [30].

Because the Digital Twin is subject to frequent changes, the question arises of how its consecutive updates (e. g. replacement for a machine in a larger hall) can be conducted without restraint to repeat the entire process described above [31]. A simple procedure can be therefore developed in which the user company is en-abled to undertake the initial part of the update process itself. For this purpose, an application is being developed with which the partial spaces of the hall to be updated in the Digital Twin can be scanned using the camera of a standard smartphone. The application will guide the user during the recording process, so that handling can be carried out quickly and easily. The scan recorded with the application is further processed automatically to a high degree and the Digital Twin is updated. With the extension of the object recognition for the part of a model to be developed, the existing Digital Twin is analysed, and changes are automatically overlayed. This can be, e. g. used to integrate behaviour models of a CNC machine or a plant.

8. DEPLOYMENT

The commercial use of the described procedure and tool for generation of Digital Twin is provided by the portal OpenDESC.com which is in service for more than 23 years. It provides service for transfer and translation of the auxiliary data (primarily CAD) in the industrial supply chain. It offers services globally and is used by more of 100 customers located worldwide [32].

A deployment schema for generation of Digital Twin is depicted in Figure 12 as a collaborative procedure where the tasks scan, object recognition and process planning/layout design are undertaken by different teams. The scan is conducted on the customer's premise, either by customer or a service provider. Then point cloud data are sent to OpenDESC.com where object recognition is done as described in section 4. Results can be transmitted to the plant engineering / process planning where the Digital Twin will be finally completed.

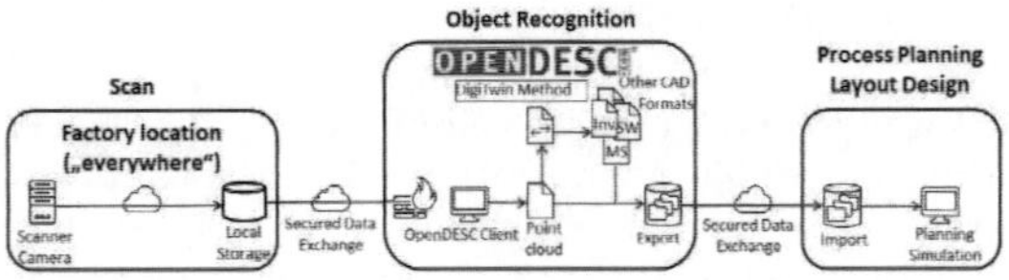

Figure 12: Workflow generation of Digital Twin as a service

9. REFERENCES

1. GRIEVES, M., VICKERS, J., 'Digital Twin: Mitigating Unpredictable, Undesirable Emergent Behavior in Complex Systems', *Transdisciplinary Perspectives on Complex Systems: New Findings and Approaches*, Springer International, Cham, 2017, pp. 85-113.

2. ERIKSSON, A., SEDELIUS, E., BERGLUND, J., JOHANSSON B., 'Virtual factory layouts from 3D laser scanning – A novel framework to define solid model requirements', *Procedia CIRP*, 2018, Vol. 76, pp. 36-41.

3. BESBES M., ZOLGHADRI M., COSTA AFFONSO R., MASMOUDI F., HADDAR M., '3D facility layout problem', *Journal of Intelligent Manufacturing*, 2020.

4. LINDSKOG, E., BERGLUND, J., VALLHAGEN, J., JOHANSSON, B., 'Layout planning and geometry analysis using 3D laser scanning in production system redesign', *Procedia CIRP*, 2016, Vol. 44, pp. 126 – 131.

5. AL-ZUBAIDI, S.Q.D., FANTONI, G., FAILLI, F., 'Analysis of drivers for solving facility layout problems: Literature review', *Journal of Industrial Information Integration*, Vol. 21, 2021, 100187.

6. QUINTANA, B., PRIETO, S.A., ADÁN, A., VÁZQUEZ, A.S., 'Semantic scan planning for indoor structural elements of buildings', *Advanced Engineering Informatics*, Vol. 30, 2016, pp. 643–659.

7. SOMMER, M. SEIFFERT, K., 'Scan Methods and Tools for Reconstruction of Built Environments as Basis for Digital Twins', *DigiTwin: An Approach for Production Process Optimization in a Built Environment*, 2021, Springer Nature.

8. SOMMER, M., STJEPANDIĆ, J., STOBRAWA, S., VON SODEN, M., 'Automated Generation of a Digital Twin in Manufacturing for a Built Environment Using Scan and Object Detection', *Journal of Industrial Information Integration*, 2021.

9. POZA-LUJAN, J.-L., POSADAS-YAGÜE, J.-L., SIMÓ-TEN, J.-E., BLANES, F., 'Distributed Architecture to Integrate Sensor Information: Object Recognition for Smart Cities', *Sensors*, 2020.

10. CARVALHO, L.E., VON WANGENHEIM, A., '3D object recognition and classification: a systematic literature review', *Pattern Analysis and Applications*, Vol. 22, 2019, pp. 1243–1292.

11. OSTROSI, E., STJEPANDIĆ, J., FUKUDA S., KURTH M., 'Modularity: New trends for product platform strategy support in concurrent engineering', *Advances in Transdisciplinary Engineering*, Vol. 1, 2014, pp. 414-423.

12. SINGH, R.D., MITTAL, A., BHATIA, R.K.., '3D convolutional neural network for object recognition: a review', *Multimedia Tools and Applications*, Vol. 78, 2019, pp. 15951–15995.

13. RANGEL, J.C., MARTÍNEZ-GÓMEZ, J., ROMERO-GONZÁLEZ, C., GARCÍA-VAREA, I., CAZORLA, M., 'Semi-supervised 3D object recognition through CNN labeling', *Applied Soft Computing Journal*, 2018, Vol. 65, pp. 603-613.

14. LIU, L., OUYANG, W., WANG, X., FIEGUTH, P., CHEN, J., LIU, X., PIETIKÄINEN, M., 'Deep Learning for Generic Object Detection: A Survey', *International Journal of Computer Vision*, Vol. 128, 2020, pp. 261–318.

15. N.N., 'Princeton Modelnet Repository', *https://modelnet.cs.princeton.edu/#*, 2021.

16. WANG, K., KIM M.-K., 'Applications of 3D point cloud data in the construction industry: A fifteen-year review from 2004 to 2018', *Advanced Engineering Informatics*, Vol. 39, 2019, pp. 306–319.

17. STJEPANDIĆ, J., SOMMER, M., 'Object Recognition Methods in a Built Environment', *DigiTwin: An Approach for Production Process Optimization in a Built Environment*, 2021, Springer Nature.

18. SOMMER, M., STJEPANDIĆ, J., STOBRAWA, S., VON SODEN, M., 'Automated Generation of a Digital Twin of a Manufacturing System by Using Scan and Convolutional Neural Networks', *Advances in Transdisciplinary Engineering*, Vol. 12, 2020, pp. 363-372.

19. DENKENA, B., DITTRICH, M.-A., STOBRAWA, S., STJEPANDIĆ, J., 'Efficient

generation of a digital twin using object detection for data acquisition and XML-interface for model creation', *Procedia CIRP*, 2020, Vol. 93, pp. 274-279.

20. SOMMER, M., STJEPANDIĆ, J., STOBRAWA, S., VON SODEN, M., 'Automatic generation of digital twin based on scanning and object recognition', *Advances in Transdisciplinary Engineering*, Vol. 10, 2019, pp. 645-654.

21. CZERNIAWSKI, T., LEITE, F., 'Automated digital modeling of existing buildings: A review of visual object recognition methods', *Automation in Construction*, 2020, Vol. 113, 103131.

22. SALEM, B., STJEPANDIĆ, J., STOBRAWA, S., 'Assessment of methods for industrial indoor object recognition', *Advances in Transdisciplinary Engineering*, Vol. 10, 2019, pp. 390-399.

23. BONDAR, S., RUPPERT, C., STJEPANDIĆ J., 'Ensuring data quality beyond change management in virtual enterprise', *International Journal of Agile Systems and Management*, Vol. 7, 2014, Nos. 3/4, pp. 304–323.

24. SOMMER, M., STJEPANDIĆ, J., STOBRAWA, S., VON SODEN, M., 'Improvement of factory planning by automated generation of a digital twin', *Advances in Transdisciplinary Engineering*, Vol. 12, 2020, pp. 453-462.

25. DENKENA, B., STOBRAWA, S., SOMMER, M., STJEPANDIĆ, J., VON SODEN, M., 'Production planning using the digital twin', *WT Werkstattstechnik*, 2020, 110(10), pp. 661-665.

26. REBOLJ, D., PUČKO, Z., ČUŠ BABIČ, N., BIZJAK, M., MONGUS, D., 'Point cloud quality requirements for Scan-vs-BIM based automated construction progress monitoring', *Automation in Construction*, Vol. 84, 2017, pp. 323–334.

27. MA, Z., LIU S., 'A review of 3D reconstruction techniques in civil engineering and their applications', *Advanced Engineering Informatics*, Vol. 37, 2018, pp. 163–174.

28. LI C.T., CHENG J.C.P., CHEN K., 'Top 10 technologies for indoor positioning on construction sites', *Automation in Construction*, 2020, Vol. 118, 103309.

29. GARCIA-GARCIA A., GARCIA-RODRIGUEZ, J., ORTS-ESCOLANO, S., OPREA, S., GOMEZ-DONOSO, F., CAZORLA, M., 'A study of the effect of noise and occlusion on the accuracy of convolutional neural networks applied to 3D object recognition', *Computer Vision and Image Understanding*, 2017, Vol. 164, pp. 124–134.

30. BONDAR, S., SALEM, B., STJEPANDIĆ., J., 'Indoor Object Reconstruction based on Acquisition by Low-cost Devices', *Advances in Transdisciplinary Engineering*, 2018, Vol. 7, pp. 113-122.

31. STJEPANDIĆ, J., SOMMER, M., STOBRAWA, S., 'Digital Twin: Conclusion and Future Perspectives', *DigiTwin: An Approach for Production Process Optimization in a Built Environment*, 2021, Springer Nature.

32. BONDAR, S., POTJEWIJD, L., STJEPANDIĆ, J. 'Globalized OEM and tier-1 processes at SKF', *Advanced Concurrent Engineering*, Springer-Verlag, London, 2012, pp. 789-800.

9. AUTHORS BIOGRAPHY

Mathias Grau holds the current position of Head of BU Marine, Offshore and Plant at PROSTEP. He is responsible for development and implementation of digital PLM in Marine, offshore and Plant industries. His previous experience includes product manager for OpemPDM Ship and various research and development activities related to Digital Master and Twin.

Josip Stjepandić holds the current position of head of business unit at PROSTEP AG. He is responsible for Digital Twin. His previous experience includes method development, engineering collaboration and knowledge-based engineering.

The Royal Institution of Naval Architects
8-9 Northumberland Street
London, WC2N 5DA
United Kingdom

ISBN 978-1-7138-8301-2

CURRAN ASSOCIATES INC.
proceedings.com

Workshop on Applications of Tree Automata in Natural Language Processing 2010

(ATANLP 2010)

Held at ACL 2010

Uppsala, Sweden
16 July 2010

ISBN: 978-1-61738-809-5